STUDENT STUDY GUIDE

David Bechtel
DeVry Institute of Technology

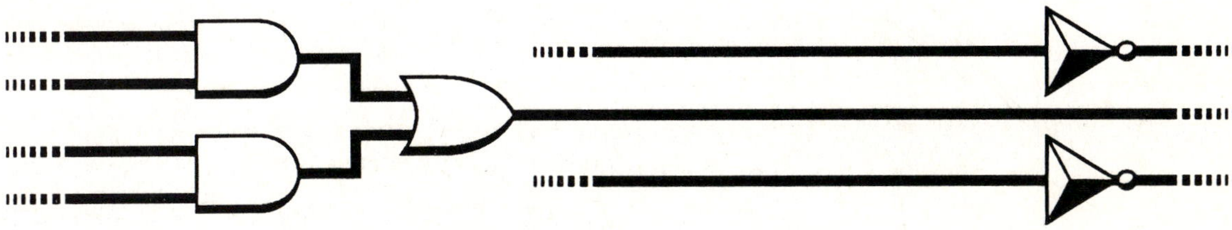

Digital Electronics

A Practical Approach

THIRD EDITION

WILLIAM KLEITZ
Tomkins Cortland Community College

REGENTS/PRENTICE HALL, Englewood Cliffs, New Jersey 07632

Editorial/Production supervision and interior design: **Marcia Krefetz**
Manufacturing buyer: **Ed O'Dougherty**
Prepress buyer: **Ilene Levy**
Acquisitions editor: **Judy Casillo**

A Division of Simon & Schuster
Englewood Cliffs, New Jersey 07632

Printed in the United States of America

10 9 8 7 6 5 4 3 2 1

ISBN 0-13-210311-7

Prentice-Hall International (UK) Limited, *London*
Prentice-Hall of Australia Pty. Limited, *Sydney*
Prentice-Hall Canada Inc., *Toronto*
Prentice-Hall Hispanoamericana, S.A., *Mexico*
Prentice-Hall of India Private Limited, *New Delhi*
Prentice-Hall of Japan, Inc., *Tokyo*
Simon & Schuster Asia Pte. Ltd., *Singapore*
Editora Prentice-Hall do Brasil, Ltda., *Rio de Janeiro*

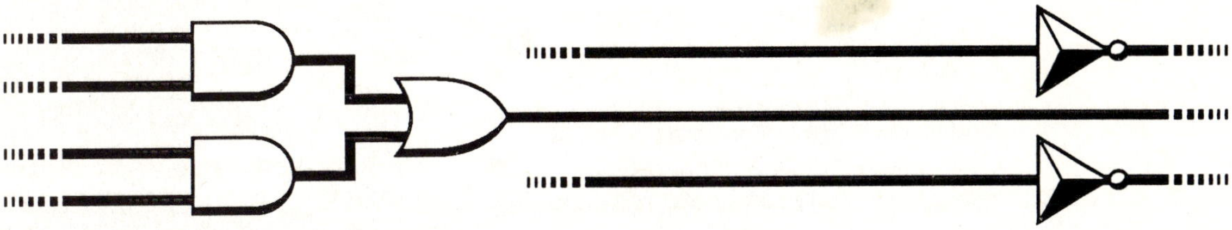

Contents

Preface

This study guide is intended as a companion to the textbook. It is hoped that the student will find it a useful tool. Each chapter in the textbook has been outlined using the same chapter titles and subtitles. The outline emphasizes the major points that the student should have gotten from his study of the textbook. The student should read over each point in the study guide as a means of reviewing the textbook and any point not clearly understood should be reread from the textbook. The student is encouraged to make use of highlighting and margin notes to bring together the points in the study guide and the reading of the textbook.

Following each chapter outline is a test that the student can use to evaluate his understanding of the material. The student is encouraged to have paper and pencil and to write out the answers to each chapter test. Then, the student's answers can be checked against the solutions found at the back of the study guide. Any question that the student misses should be looked up in the textbook and re-studied by the student.

1

CHAPTER

Number Systems and Codes

Digital Representations of Analog Quantities

- Many natural physical quantities are analog in nature.
- Most electronic machines, on the other hand, are digital.
- For most applications the continuous analog signal must be converted to a digital signal using a series of ON and OFF voltages.
- The audio recording of music is an example of how an analog signal can be converted to a digital signal and then back to an analog signal again.

Digital Versus Analog

- Digital systems use discrete digits to represent numbers. Everything can be represented by a series of 1s and 0s.
- Analog systems use continuous varying physical quantities. Usually, a voltage or current is used to represent a specific value.

Decimal Number System

- The decimal system uses the number ten as a base.
- The decimal system has ten digits: 0, 1, 2, 3, 4, 5, 6, 7, 8 and 9.
- Each number position represents a weighting factor which is a power of the base ten. (1, 10, 100, 1000, etc.)

Binary Number System

- The binary system uses the number two as a base.
- The binary system has only two digits, 0 and 1; which make it useful to represent switch positions (open or closed).
- Each binary position contains one binary digit or bit.
- The leftmost bit position is called the most significant bit (MSB).
- The rightmost bit position is called the least significant bit (LSB).
- Each number position represents a weighting factor which is a power of the base two. (1, 2, 4, 8, 16, 32, 64, etc.)
- Fractional binary numbers have weighted values that are developed by repeated division by 2. (0.5, 0.25, 0.125, etc.)
- To convert binary to decimal, write down the power of two that represents each 1 in the binary number and add up these powers to get the decimal number.

Decimal-to-Binary Conversion

- Sum-of-Weights Method

  Find all the powers of 2 that fit into a number starting with the largest value.
  Write the binary number with 1s in the bit positions of these weights.
  The sum of these weights should equal the decimal number.
- Successive Division Method

  Repeatedly divide the number by the base 2.
  The remainders from each division represent the binary number.
  The first division produces the MSB and the last division produces the LSB.

Octal Numbering System

- The octal system uses the number eight as the base.
- The octal system uses only eight digits: 0, 1, 2, 3, 4, 5, 6, and 7.
- Each octal digit position represents a different power or weight of the base eight. (1, 8, 64, 512, etc.)
- Octal is primarily used as a "shorthand" means of representing binary numbers.
- The octal system makes it much easier for an operator to enter or read binary numbers.

Octal Conversions

- Binary to Octal

  Group the binary digits in groups of three starting with the least significant bit.
  Write down the octal digit that represents each group of three binary digits.
- Octal to Binary

  Write down the binary group of three bits for each octal digit.

- Octal to Decimal
  Multiply each octal digit by its weighting factor (power of the base 8).
  Add these products together to get the decimal number.
- Decimal to Octal
  Successively divide the decimal number by 8.
  The remainders are the octal digits starting with the most significant digit first.

Hexadecimal Numbering System

- The hexadecimal system uses the number sixteen as a base.
- The hexadecimal system has 16 digits: 0, 1, 2, 3, 4, 5, 6, 7, 8, 9, A, B, C, D, E, and F.
- The hexadecimal digits A thru F are equal to the decimal values 10 thru 15.
- Each hexadecimal position represents a different weighting or power of the base 16. (1, 16, 256, 4096, etc.)
- The hexadecimal system, like the octal system, is a "shorthand" means of representing binary numbers.
- Two hexadecimal digits are used to represent a binary byte and a single hexadecimal digit represents a binary nibble.

Hexadecimal Conversions

- Binary to Hexadecimal
  Group the binary bits in groups of four starting with the least significant bit.
  Write down the hexadecimal digit that represents each binary group of four.
- Hexadecimal to Binary
  Write the binary group of four bits that represents each hexadecimal digit.
- Hexadecimal to Decimal
  Multiply each hexadecimal digit by its weighting factor (power of the base 16). Add these products together to get the decimal number.
  An additional method is to convert the decimal number to its binary value first; then, convert the binary number to hexadecimal.
- Decimal to Hexadecimal
  Successively divide the decimal number by 16.
  The remainders are the hexadecimal digits starting with the most significant digit first.
  If the remainder is 10 or greater it must be converted to its hexadecimal letter symbol (A, B, C, D, E, F).

Binary-Coded-Decimal System

- Binary-Coded-Decimal (BCD) is a shorthand method for representing decimal numbers in binary.
- Each decimal digit is coded into a 4-bit binary code.
- To convert decimal numbers to BCD simply write down the four bit binary code for each decimal digit.

- To convert BCD to decimal, group binary into groups of 4 and convert each group into its decimal digit.
- There are six binary codes that are illegal in the BCD system.

Comparison of Number Systems

- The Octal, Hexadecimal and BCD systems are useful because they can be converted from binary one digit at a time.
- The Octal system uses groups of three binary bits and the Hexadecimal and BCD systems use groups of four binary bits.
- Decimal is not as easily converted to Binary.
  The entire decimal number must be converted by the Sum-of-Weights method or the Successive Division method.

The ASCII Code

- American Standard Code for Information Interchange (ASCII) pronounced "askee".
- The code is used to represent letters, symbols and numbers.
- It is a 7-bit code used by computers to input and output information.
- The code is divided into a group of 3 bits (MSD) and a group of 4 bits (LSD).

Applications of the Number Systems

- Binary numbers are often used to represent physical quantities rather than numeric values.
- Binary numbers are often described with their hexadecimal value for easy handling.
- The technician should be able to translate between binary and hexadecimal with ease.
- In some cases octal numbers are used to represent binary values.
- In cases where information must be displayed, decimal values are often coded in BCD.
- For more detailed information, numbers, letters and special symbols are coded as ASCII codes.

Additional Notes

The Octal, Hexadecimal, and BCD systems have an advantage over the Decimal system in that they are single-digit conversion systems. This means that they can be converted to binary one digit at a time. Decimal, on the other hand, must be converted by manipulation of the entire number. This can be done with the Sum-of-Weights method or the Successive Division method. The Octal, Hexadecimal, and BCD systems provide a much faster and easier way of representing binary numbers than converting them to decimal.

The Octal and Hexadecimal systems are direct number conversions. That is, they are the same numerical value as the binary number that they represent. If a binary number is converted to decimal and its hexadecimal representation is also converted to decimal, they will both give the same decimal number. On the other hand, the BCD and ASCII systems are coding systems. In a code, a letter or number has a binary sequence that is used to represent that letter or number. The sequence does not necessarily equal the numerical value of the number.

TEST 1

Convert the following binary numbers to decimal.

1. 01011_2

2. 10011010_2

Convert the following binary numbers to octal.

3. 001101011_2

4. 010111100_2

Convert the following binary numbers to hexadecimal.

5. 01011101.0101_2

6. 11001011.110_2

Convert the following decimal numbers to 8-bit binary

7. 35

8. 187

Convert the following octal numbers to decimal.

9. 17_8

10. 35_8

Convert the following decimal numbers to hexadecimal.

11. 125

12. 74

Convert the following octal numbers to binary.

13. 76_8

14. 104_8

Convert the following hexadecimal numbers to decimal.

15. $B5_{16}$

16. $1CF_{16}$

Convert the following decimal numbers to octal.

17. 281

18. 39

Convert the following hexadecimal numbers to binary.

19. $C9_{16}$

20. $14B_{16}$

Convert the following BCD numbers to decimal.

21. 010101101001_{bcd}

22. 100000000011_{bcd}

Convert the following decimal numbers to BCD.

23. 127

24. 469

25. Decode the following ASCII message.

    1010011101010010101011000100101100101000001001000100000110100101000100

2
CHAPTER

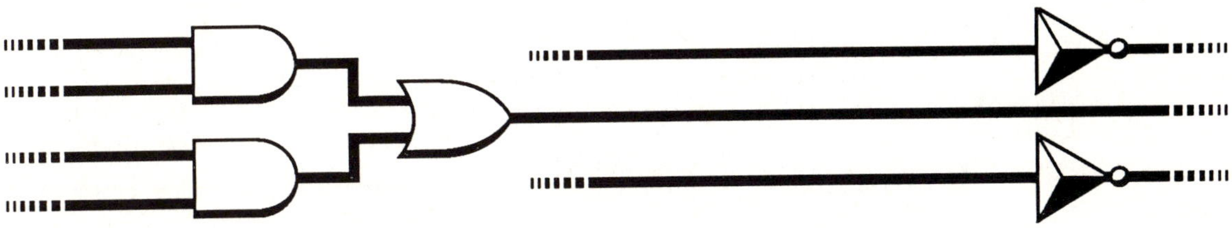

Digital Electronic Signals and Switches

Digital Signals

- Logic states (1's and 0's) are represented by voltage levels (usually 1 = 5v and 0 = 0v).
- A timing diagram is a plot of voltage versus time.
- The timing diagram shows the relationship between the digital waveforms as they vary relative to time.
- Digital circuits respond to the binary level of the input signal and not necessarily to the actual input voltage.

Clock Waveform Timing

- Special clock and timing circuits are used to produce periodic clock waveforms.
- The period of a waveform is the length of time of one complete cycle of the waveform.
- The frequency of a waveform is determined by calculating the reciprocal of the period of the waveform.
- The frequency is measured in Hertz (Hz.) and the period is measured in Seconds (Sec.).

Serial Representation

- The serial format transmits binary information one bit at a time over a single pair of conductors.
- The serial format is inexpensive but slow.
- The serial format transmits the LSB first and each bit occupies a separate clock pulse.

Parallel Representation

- The parallel format provides a separate conductor (or pair of conductors) for each digit of the binary word to be transmitted.
- The parallel format is more expensive than the serial format, but much faster.
- The entire data word is transmitted in one clock pulse.

Switches in Electronic Circuits

- Digital logic levels can be caused by manual switches or electromechanical relays or by use of semiconductor devices such as transistors and diodes.
- Ideal switches have 0 ohms of resistance when they are closed and infinite resistance when they are open.
- A SPDT (Single-Pole-Double-Throw) switch can be used to connect an output terminal to 5v (logic 1) or 0v (logic 0).

A Relay as a Switch

- An electromechanical relay has switch contacts that are opened or closed by an external voltage.
- Energizing the relay coil may open some contacts (normally closed) and may close other contacts (normally open).
- Relays are used to provide total isolation between the triggering source and the output.
- Relays are also used to switch very high currents.
- Relays are difficult to use because they require large currents to energize the coil and they operate very slowly when compared to semiconductor devices.
- A voltage divider can be used on the output terminals of a relay to set-up the required voltage levels.
- A clock circuit can be used to drive a relay and the output can be compared to the clock with a timing diagram.

A Diode as a Switch

- A diode is a semiconductor device that will allow current to flow in one direction, but not in the other.
- A diode can act as an open switch (reverse bias) or as a closed switch (forward bias).
- A forward biased silicon diode has a 0.7v "cut-in voltage" that must be exceeded before the diode will conduct.

A Transistor as a Switch

- The three terminals of the transistor are the emitter, base and collector.
- Transistors can be manufactured as either NPN type or PNP type. The middle letter indicates the base material.
- When the proper polarity of voltage is applied to the base of a transistor, the transistor will short from emitter to collector (closed switch).
- The reverse polarity or no voltage on the base will cause the transistor to open from emitter to collector (open switch).
- To turn on (short) a NPN transistor the base voltage must be positive and for a PNP transistor the base voltage must be negative.
- To analyze the output of a transistor switch, consider that the emitter-collector connection is either an open circuit or a short circuit.

The TTL Integrated Circuit

- One of the most popular of integrated circuit logic families is the Transistor-Transistor Logic (TTL).
- The basic function of a TTL circuit is that of an inverter.
- The output of an inverter is the opposite logic state from that on the input.
- The output voltage of a single transistor inverter is dependant upon the ratio of the collector resistor and the load resistor. The performance of the output circuit can be greatly improved by using two transistors in the output circuit.
- TTL circuits use a totem-pole arrangement on the output (when one transistor is ON, the other one is OFF).
- Most TTL IC's come in a Dual-in-line package (DIP).
- DIP circuits usually have several of the same logic functions inside one package.

The CMOS Integrated Circuit

- CMOS (Complementary-Metal-Oxide Semiconductor) is another very popular family of digital integrated circuits.
- The advantage of CMOS is the very low amount of power required to operate the logic.

- The problem with using CMOS is that it operates at slower speeds than TTL.
- CMOS can also be damaged by static charges if improperly handled.

Additional Notes

There are many different families of digital logic. The most popular of these are TTL and CMOS. Other popular families include ECL (Emitter Coupled Logic), MOS (Metal-Oxide Semiconductor), and I^2L (Integrated Injection Logic). There are many reasons why one family may be used in a given situation over other families. Some of the considerations used to compare logic families include:

1. Switching Speed - The time it takes the output of a circuit to respond to a change on the input.
2. Power Dissipation - The amount of power drawn by the circuit from the power supply. This power is usually dissipated in the form of heat.
3. Noise Margin - The difference in voltage between the output logic level and the required input voltage of the next stage. The noise margin is a measure of a circuit's immunity to false triggering by an outside source.
4. Fanout - The number of circuits that can be driven by a single output.
5. Cost - The cost can be expressed as cost per chip or, more likely, as cost per logic gate on that chip. A 7404 chip (TTL hex inverter) that costs 24 cents for the chip costs only 4 cents per inverter circuit.

TEST 2

1. What is the frequency of a clock waveform if the period of that waveform is 1.25 usec?

2. What is the period of a clock waveform that has a frequency of 15.4 kHz?

3. What is it that is used to graphically show the time relationship between two or more digital waveforms?

4. How long will it take to serially transmit the hexadecimal number $C3_{16}$ if the clock frequency is 1.5 MHz?

5. The hexadecimal number $4B_{16}$ is transmitted as a 8-bit word in parallel. What is the time required for this transmission if the clock frequency is 2.25 MHz?

6. Sketch the serial representation of the hexadecimal number $B4_{16}$.

7. Sketch the 4-bit parallel representation of the hexadecimal number $3A_{16}$.

8. What is the advantage of using serial data transmission over parallel data transmission?

9. Why is parallel data transmission preferred over serial data transmission for most applications?

10. Sketch the voltage V_{out} for the circuit shown in Figure 2-1.

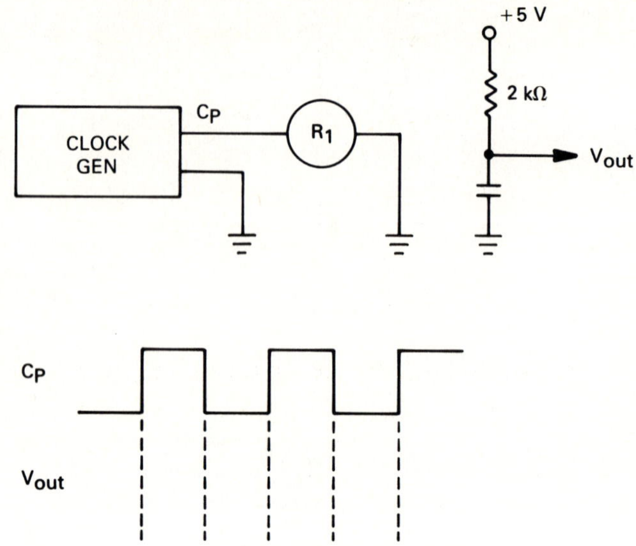

Figure 2-1

11. Draw a timing diagram for the circuit shown in Figure 2-2.

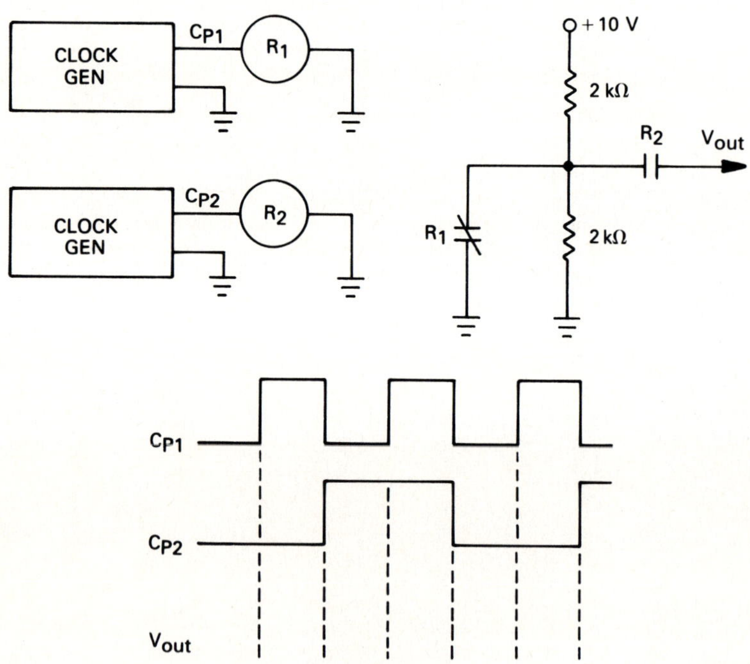

Figure 2-2

12. Determine if diodes D1 and D2 in Figure 2-3 are forward or reverse biased.

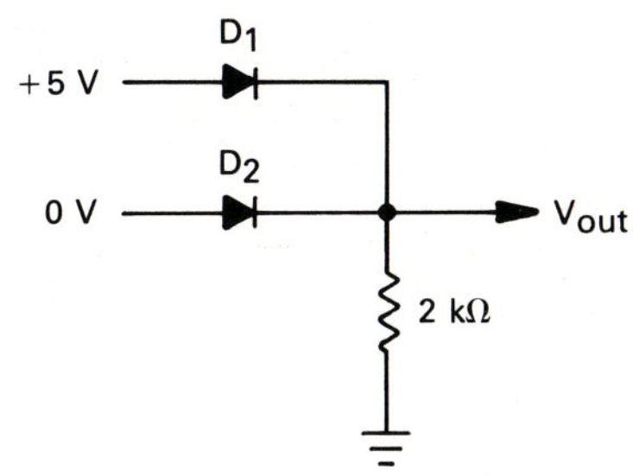

Figure 2-3

13. Determine the output voltage V_{out} for the circuit of Figure 2-3.

14. If the input voltage on D1 in Figure 2-3 is changed to 0v, what is the output voltage V_{out}?

15. Determine the voltage on the collector V_c of the transistor circuit shown in Figure 2-4.

Figure 2-4

16. Is the transistor in Figure 2-4 considered ON or OFF?

17. If the input voltage on the base of the transistor in Figure 2-4 is changed to +10v, what is the new voltage at the collector V_c?

18. Why, in most applications, are transistor switches used in place of relays?

19. What is the output voltage V_{out} of the circuit in Figure 2-5?

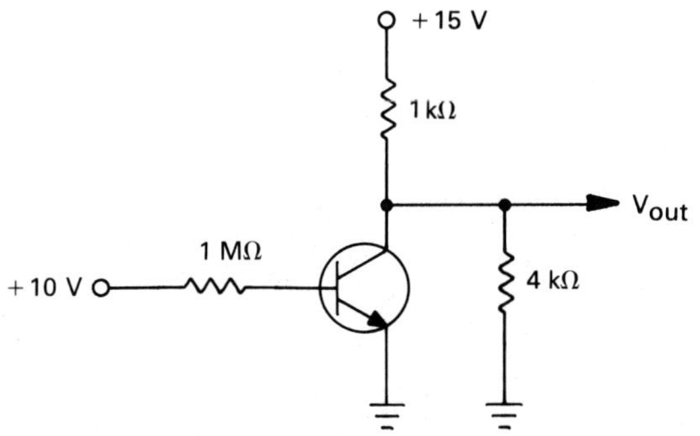

Figure 2-5

20. Recalculate the output voltage V_{out} for the circuit of Figure 2-5 if the input voltage is changed to 0v.

21. Sketch the output voltage for Figure 2-5 if the input is a clock signal that pulses from 0v to +10v.

22. Why is the circuit of Figure 2-5 called an inverter?

23. Why does the TTL family use a totem pole circuit on the output?

24. What is the advantage of using CMOS logic over TTL?

25. What is a disadvantage of CMOS in place of TTL?

3

CHAPTER

Basic Logic Gates

The AND Gate

- The output of an AND gate is true (logic 1) if and only if all of the inputs to the gate are true (logic 1).
- A truth table is used to illustrate how the output of a gate responds to all possible combinations on the inputs to the gate.
- The AND gate can be illustrated with a series connection of manual switches or transistor switches.
- The AND function can be expressed mathematically with the Boolean Expression $X = A \bullet B$.
- The standard multiplication symbol is used in Boolean Algebra to represent the AND function.
- The number of combinations of a truth table is equal to 2^N where N is the number of inputs.

The OR Gate

- The output of an OR gate is true (logic 1) if any or all of the inputs are true (logic 1).
- The OR gate can be illustrated with a parallel connection of manual switches or transistor switches.
- The Boolean Expression for a two input OR gate is $X = A + B$.
- The standard addition symbol (+) is used in Boolean Algebra to represent the OR function.
- The only time the output of an OR gate is low is when all the inputs are low.

Timing Analysis

- A timing diagram is a graph of the output of a logic gate with respect to the inputs of the gate.
- A timing diagram plots voltage (vertical) with respect to time (horizontal).
- The timing can also be seen as waveforms on an oscilloscope or on a logic analyzer.
- In order to determine the proper output waveform from a logic gate, simply divide the input diagrams into time segments where the inputs are constant and determine the state of the output (high or low) for that segment from the truth table.
- If the situation comes up where it does not make any difference which state an input is in (either way the output does not change), the input is said to be in a don't care condition.

Enable and Disable Functions

- AND and OR gates can both be used to enable or disable a transmitted waveform.
- For a two input AND gate, one input is the signal and the other input is the enable pulse.
- The enable of an AND gate is high active. That is, when the enable is high the input signal will appear on the output.
- When the AND gate enable input is low, the output will remain a constant low signal.
- A two input OR gate can also be used with one input the desired signal and the other input is the enable.
- The enable input of an OR gate is low active. This means that the output will be a copy of the input signal when the enable is low.
- When the enable input of an OR gate is high, the output of the gate will be a constant high signal.

Using Integrated-Circuit Logic Gates

- Many different types of logic gates are available on integrated circuits (ICs).
- The information about these circuits along with their pin assignments can be found in the manufacturers manual.
- The terms quad (four), triple (three) and dual (two) are used to indicate the number of logic gates on an IC.
- These logic gates can usually be obtained in a 14-pin Dual-in-Line Package (DIP) IC where pin 14 is $+V_{cc}$ and pin 7 is ground.
- Pin 1 is identified by a small circle next to it or by a notch in the end of the case between pins 1 and 14.

Introduction to Troubleshooting Techniques

- Troubleshooting is the steps used to locate the fault or trouble in a circuit.
- The first step in troubleshooting is to understand how a particular IC is supposed to work.
- One tool for digital troubleshooting is the logic probe. The logic probe is used to indicate the High (1), Low (0), or floating (open circuit) condition of any pin on a digital IC.
- The second tool used in digital troubleshooting is the logic pulser. The pulser is used to inject a series of High and Low pulse signals into a logic gate.
- To test an AND gate, connect all inputs but one high. Connect the remaining input to the pulser and check the output with the probe. The output should be pulsing.
- To test an OR gate, connect all inputs except one low. Connect the unused input to the pulser and check the output with the probe. The output should again be pulsing.
- An experienced technician can use visual inspection as a troubleshooting tool. The technician will look for conditions such as misaligned or broken IC pins, cracked circuit boards, solder bridges and burnt or overheated components.

Additional Notes

The AND gate and the OR gate are basic building blocks that will be used to construct more complex logic functions. The OR gate is sometimes called the "Either/Or Both" gate and the AND gate is sometimes called the "Coincidence" gate. These two gates, when combined with the NOT gate, can be used to construct about any logic function desirable.

The output waveform can most easily be determined if the input signals are first broken up into time segments where in each time segment the inputs are constant. Whenever an input changes, mark another time segment. Then, for each time segment determine the state of the output from the truth table for that logic gate. If this is repeated for each time segment then the result should be a continuous waveform on the output.

TEST 3

1. Construct a truth table for a three input AND gate with inputs A, B and C and an output X.

2. Construct a truth table for a four input OR gate with inputs A, B, C and D and an output Y.

3. Write the Boolean Expression for the three input AND gate of problem 1.

4. Write the Boolean Expression for the four input OR gate of problem 2.

5. How many input combinations would a truth table have for a six input AND gate?

6. Sketch the output waveform X for the logic gate and input waveforms shown in Figure 3-1.

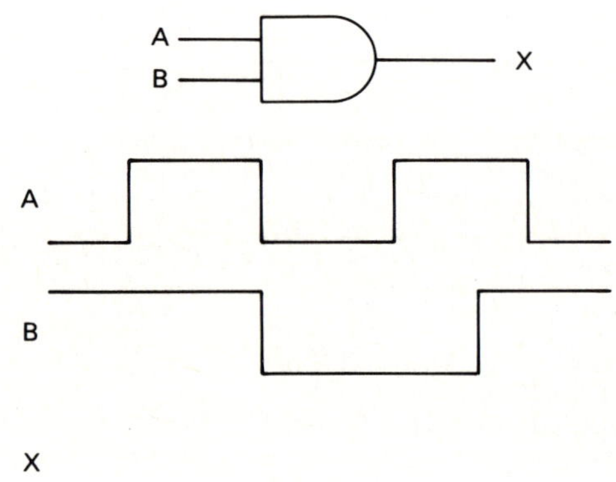

Figure 3-1

7. Sketch the output waveform Y for the logic gate and input waveforms shown in Figure 3-2.

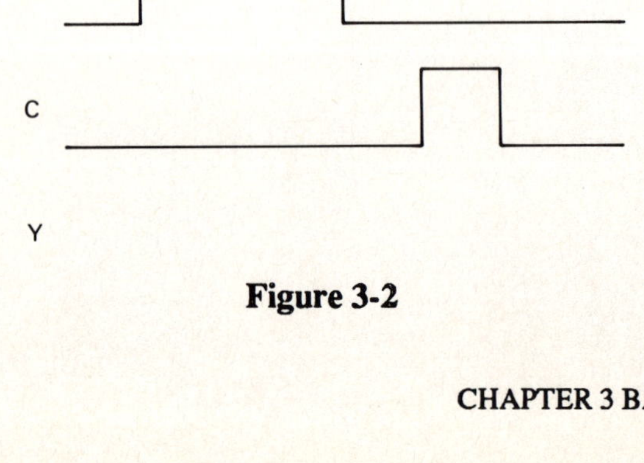

Figure 3-2

8. Sketch the input signal on input C that will give the required output of the logic gate of Figure 3-3.

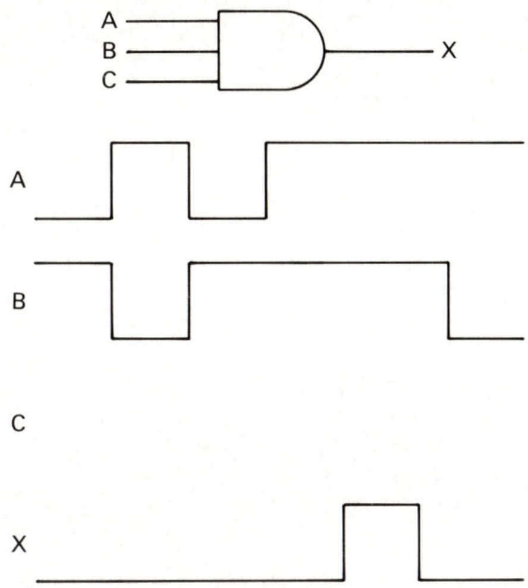

Figure 3-3

9. How many inputs of a four input AND gate must be high in order for the output of the logic gate to go high?

10. If the output of a three input OR gate must be a logic low, what must the condition of the inputs be?

11. A clock signal with a period of 1 usec. is applied to the input of an enable gate. The output must contain six pulses. How long must the enable Pulse be active?

12. If the output of the enable circuit in problem 11 must be high when the circuit is disabled, should an AND gate or an OR gate be used for the circuit?

13. Will the enable circuit defined in problem 11 and 12 have an active high or an active low enable?

14. How many AND gates are found in a 7411 IC?

15. How many inputs are on the logic gates of a 74HC21 IC?

16. What are the pin numbers of the outputs of the gates in a 7432 IC?

17. How is a logic probe used to test a digital circuit?

18. What is the function of a logic pulser in testing digital ICs?

19. A logic probe is used to test the pins of a 7411 IC with the following results. Is there a problem with the chip and if so, what is the problem?

pin	indicator	pin	indicator
1	ON	14	ON
2	PULSING	13	ON
3	PULSING	12	PULSING
4	ON	11	ON
5	ON	10	OFF
6	OFF	9	ON
7	OFF	8	OFF

20. A logic probe is again applied to the pins of a 7421 IC with the following results. Is there a problem with this circuit and if so, what is the problem?

pin	indicator	pin	indicator
1	ON	14	ON
2	PULSING	13	ON
3	DIM	12	ON
4	ON	11	DIM
5	ON	10	OFF
6	PULSING	9	PULSING
7	OFF	8	OFF

4

CHAPTER

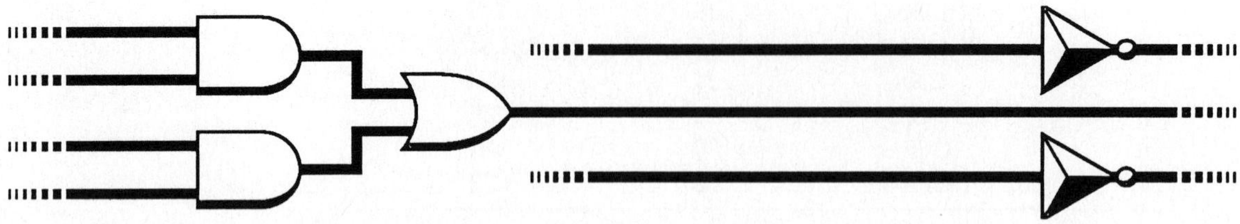

Inverting Logic Gates

The Inverter

- The output of an inverter is the complement (opposite) of the input.
- When the input to an inverter is high (1) the output is low (0); and when the input is low, the output is high.
- In Boolean Algebra the inverter operation is shown by placing a bar over the variable.

$X = \overline{A}$ (read as X equal to NOT A).

- The waveform on the output of an inverter would look like the exact opposite of the waveform on the input.
- The inverter is also often called a NOT gate.

The NAND Gate

- The NAND gate is the same function as an AND gate with the output inverted.
- The logic symbol for a NAND gate is the same as an AND gate except it has a small bubble on the output to indicate that the output is inverted.
- The Boolean Expression for the NAND gate has all input variables ANDed together and a bar over the entire expression.

$(X = \overline{ABC})$

- The truth table for the NAND gate shows the output to be just the inverse of the output of an AND gate.
(The only time the output is low is when all the inputs are high.)

- The output of a NAND gate can be shown with a timing diagram in the same manner that the output of the AND and OR gate were developed.

The NOR Gate

- The NOR gate is the same as an OR gate with the output inverted.
- The NOR gate logic symbol is an OR gate with a bubble on the output to indicate an inverted output.
- The NOR Boolean Expression has the inputs ORed together with a bar over the entire expression.

$$(X = \overline{A + B + C})$$

- The NOR gate truth table is the OR gate truth table with the output inverted. (The only time the output is high is when all the inputs are low.)
- The output of a NOR gate can be demonstrated with a timing diagram. The output is developed one segment at a time as the inputs change. The output again will follow the truth table.

Logic Gate Waveform Generation

- One type of waveform generator circuit is the Johnson Shift Counter.
- The Johnson Counter has four different output waveforms plus the complement of each. (total of 8 outputs)
- Additional logic gates can be connected to the Johnson Counter to obtain any desired waveform pattern.
- Even very specialized waveforms can be generated if the proper combination of logic gates is applied to the Johnson Counter.

Using Integrated-Circuit Logic Gates

- All logic gates are available in both TTL and CMOS logic families.
- All the gates are available in configurations of from two inputs per gate up to eight inputs per gate.
- A TTL or CMOS manual should be consulted for proper circuit configuration and pin assignment.
- High speed CMOS (74HC_ _ series) have the same pin assignments as the TTL series.

Summary of Basic Logic Gate and IEEE/IEC Standard Symbols

- The three basic logic gates are the AND, OR and the Inverter.
- The NAND gate is a combination of an AND gate followed by an inverter.
- The NOR gate is a combination of an OR gate followed by an inverter.

- There is a new IEEE/IEC standard for logic symbols that allows the reader to determine the logic function simply by interpreting the notions on the symbol.

Additional Notes

The NAND and the NOR logic gates are sometimes called the universal logic gates because the three basic building blocks of all logic (AND, OR and Inverter) can be accomplished using only NAND gates or using only NOR gates. All complex logic functions can be achieved using AND, OR and Inverter gates. If NAND and NOR gates are universal, then all complex functions can be accomplished using only NAND gates or using only NOR gates. This makes the NAND gate and the NOR gate very powerful gates. For this reason, many logic families will use a large number of NAND gates or a large number of NOR gates. The TTL logic family, for example, has a large number of the available circuits that are NAND logic gates.

TEST 4

1. Sketch the logic symbol and complete a truth table for an inverter logic gate.

2. Write the Boolean Expression for an inverter logic gate with input C and output Y.

3. Sketch the output waveform of the inverter shown in Figure 4-1.

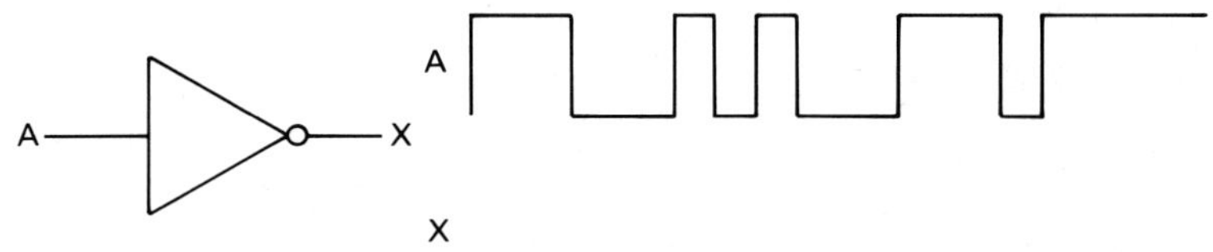

Figure 4-1

4. Write a Boolean Expression for a three input NAND logic gate with inputs A, B and C and output X.

5. What does the small bubble on the output of NAND gate logic symbol mean?

6. Complete a truth table for a two input NAND gate.

7. How many entries would a truth table for a four input NAND gate have?

8. Sketch the output waveform for the NAND logic gate shown in Figure 4-2.

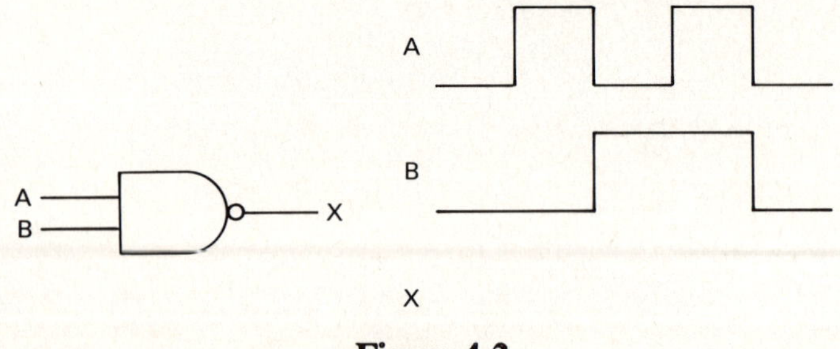

Figure 4-2

9. The NOR logic gate is the same as the operation of the ______ gate with an inverter connected to the output.

10. From the truth table for a three input NOR gate, what is the only condition of inputs A, B and C that will make the output X high?

11. Write the Boolean Expression for a two input NOR gate with inputs A and B and output Y.

12. Sketch the output Y of the three input NOR gate shown in Figure 4-3.

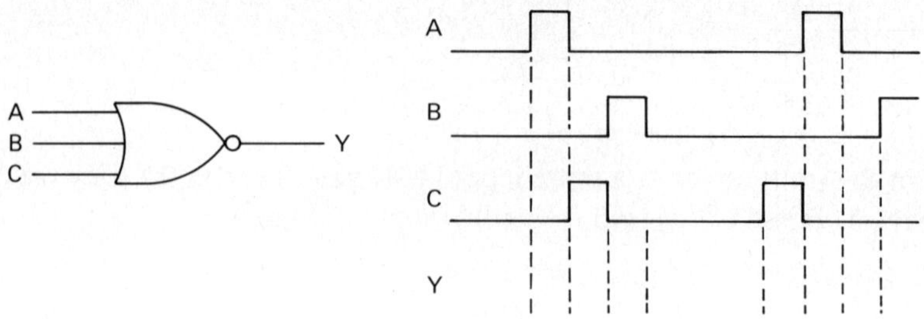

Figure 4-3

13. What is the name of a digital circuit that produces several repetitive digital waveforms?

14. Which Johnson Counter outputs must be connected to a three input AND gate to enable just the Cp 6 pulse to the output?

15. Sketch the output waveform that will result from inputting $\overline{B}$ and C into a two input NOR gate.

16. When reading a Boolean Expression, what does the word "NOT" indicate?

17. What is the circuit number of the IC that contains 4 two input AND gates in standard TTL?

18. When used with an IC, what does the term "QUAD" indicate?

19. How many pins does the 4049 IC have?

20. What is the difference between a 7400 and 7410 IC?

5

CHAPTER

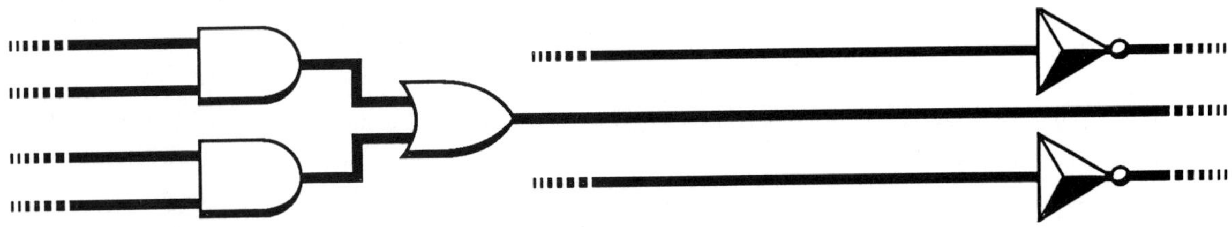

Boolean Algebra and Reduction Techniques

Combinational Logic

- Several of the basic logic gates are used to form a more complex function with combinational logic.
- A Boolean equation can be used to describe any combinational logic circuit. The Boolean equation is written in a form that will satisfy the problem.
- A Boolean reduction is used to simplify a combinational logic by using Boolean Algebra.
- Boolean Algebra can use parentheses to set off a group of variables that are ANDed to a common variable or group of variables. AB + AC = A(B + C).

Boolean Algebra Laws and Rules

- There are three laws of Boolean Algebra that are the same as ordinary algebra.
  1. The Commutative Law
     of addition A + B = B + A
     of multiplication AB = BA
  2. The Associative Law
     of addition A + (B + C) = (A + B) + C
     of multiplication A(BC) = (AB)C
  3. The Distributive Law
     A(B + C) = AB + AC
     (A + B) (C + D) = AC + AD + BC + BD

- There are ten basic rules of Boolean Algebra.

1. Anything ANDed with a 0 is equal to 0

$$A \bullet 0 = 0$$

2. Anything ANDed with a 1 is equal to itself

$$A \bullet 1 = A$$

3. Anything ORed with a 0 is equal to itself

$$A + 0 = A$$

4. Anything ORed with a 1 is equal to 1

$$A + 1 = 1$$

5. Anything ANDed with itself is equal to itself

$$A \bullet A = A$$

6. Anything ORed with itself is equal to itself

$$A + A = A$$

7. Anything ANDed with its own complement equals 0

$$A \bullet \overline{A} = 0$$

8. Anything ORed with its own complement equals 1

$$A + \overline{A} = 1$$

9. Anything complemented twice is equal to the original

$$\overline{\overline{A}} = A$$

10. The two variable rule

$$A + \overline{A}B = A + B$$
$$\overline{A} + AB = \overline{A} + B$$
$$A + AB = A$$

Simplification of Combinational Logic Circuits Using Boolean Algebra

- Complex combinational logic circuits must be reduced without changing the function of the circuit.
- Reduction of a logic circuit means the same logic function with fewer gates and/or inputs.

- The first step to reducing a logic circuit is to write the Boolean Equation for the logic function.
- The next step is to apply as many rules and laws as possible in order to decrease the number of terms and variables in the expression.
- To apply the rules of Boolean Algebra it is often helpful to first remove any parentheses or brackets.
- After removal of the parentheses, common terms or factors may be removed leaving terms that can be reduced by the rules of Boolean Algebra.
- The final step is to draw the logic diagram for the reduced Boolean Expression.

DeMorgan's Theorem

- DeMorgan's theorem allows large bars in a Boolean Expression to be broken up into smaller bars over individual variables.
- DeMorgan's theorem says that a large bar over several variables can be broken between the variables if the sign between the variables is changed.

$$\overline{A \bullet B \bullet C} = \overline{A} + \overline{B} + \overline{C}$$

$$\overline{A + B + C} = \overline{A} \bullet \overline{B} \bullet \overline{C}$$

- DeMorgan's theorem can be used to prove that a NAND gate is equal to an OR gate with inverted inputs.
- DeMorgan's theorem can be used to prove that a NOR gate is equal to an AND gate with inverted inputs.
- In order to reduce expressions with large bars, the bars must first be broken up. This means that in some cases, the first step in reducing an expression is to use DeMorgan's theorem.
- Bubble pushing is a technique to apply DeMorgan's theorem directly to the logic diagram.
  1. Change the logic gate (AND to OR and OR to AND).
  2. Add bubbles to the inputs and outputs where there were none, and remove the original bubbles.
- Logic gates can be DeMorganized so that bubbles appear on inputs or outputs in order to satisfy signal conditions rather than specific logic functions. An active-low signal should be connected to a bubble on the input of a logic gate.

The Universal Capability of NAND and NOR Gates

- NAND and NOR gates are universal logic gates.
- The AND, OR, NOR and Inverter functions can all be performed using only NAND gates.
- The AND, OR, NAND and Inverter functions can all be performed using only NOR gates.
- An inverter can be made from a NAND or a NOR by connecting all inputs of the gate together.
- If the output of a NAND gate is inverted, it becomes an AND function.
- If the output of a NOR gate is inverted, it becomes an OR function.

- If the inputs to a NAND gate are inverted, the gate becomes an OR function.
- If the inputs to a NOR gate are inverted, the gate becomes an AND function.
- When NAND gates are used to make the OR function and the output is inverted, the function becomes NOR.
- When NOR gates are used to make the AND function and the output is inverted, the function becomes NAND.

AND-OR-Invert Gates for Implementing Sum-of-Products Expressions

- Most Boolean reductions result in a Product-of-Sums (POS) expression or a Sum-of-Products (SOP) expression.
- The Sum-of-Products means the variables are ANDed to form a term and the terms are ORed. X = AB + CD.
- The Product-of-Sums means the variables are ORed to form a term and the terms are ANDed. X = (A + B) (C + D)
- AND-OR-Inverter gate combinations (AOI) are available in standard ICs and can be used to implement SOP expressions.
- The 74LS54 is a commonly used AOI.
- Programmable Logic Devices (PLDs) are available for larger and more complex functions than can be accomplished with an AOI.

Karnaugh Mapping

- Karnaugh mapping is a graphic technique for reducing a Sum-of-Products (SOP) expression to its minimum form.
- Two, three and four variable k-maps will have 4, 8 and 16 cells respectively.
- Each cell of the k-map corresponds to a particular combination of the input variable and between adjacent cells only one variable is allowed to change.
- Use the following steps to reduce an expression using a k-map.
  1. Use the rules of Boolean Algebra to change the expression to a SOP expression.
  2. Mark each term of the SOP expression in the correct cell of the k-map.
  3. Circle adjacent cells in groups of 2, 4 or 8 making the circles as large as possible.
  4. Write a term for each circle in a final SOP expression. The variables in a term are the ones that remain constant across a circle.
- The cells of a k-map are continuous left-to-right and top-to-bottom. The wraparound feature can be used to draw the circles as large as possible.
- When a variable does not appear in the original equation, the equation must be plotted so that all combinations of the missing variable(s) are covered.

System Design Applications

- The first step in any design problem is to express the problem in terms of a Boolean Expression. This can be done either from the requirements of the problem directly or from a truth table.
- The next step is to reduce the expression to simplest terms and to put it in a desired format. This step may include the use of Boolean Algebra, k-maps or both.
- The third step is to implement the function by drawing the logic using the available circuits.

Additional Notes

In some cases the question arises as to the order of operations. If an AND and an OR appear in the same expression, which is to be done first? The order of operations of Boolean Algebra are the same as standard algebra. The AND operation (same as multiplication) is performed first. Of course, parentheses can be used to alter the order of operations just as in standard algebra. In the expression AB + C, A is ANDed to B then ORed with C. In the expression, A(B + C), B is ORed with C first, then ANDed with A.

A function inside a parentheses must be accomplished first before any functions outside the parentheses. A bar over several variables can also act as a parentheses. Any function under the bar must be done before any functions not under the bar.

TEST 5

1. Write a Boolean Expression for the logic diagram shown in Figure 5-1.

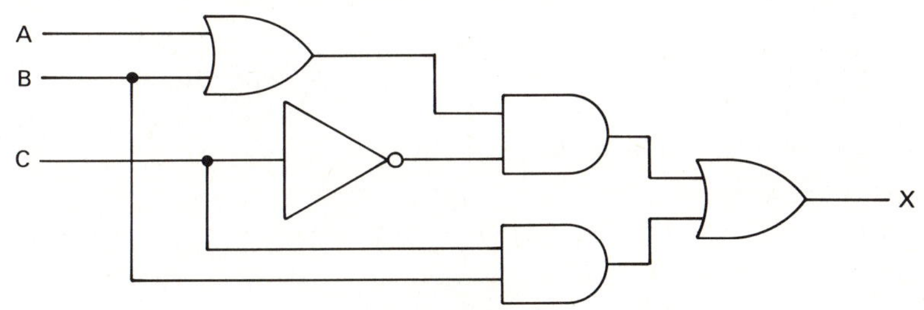

Figure 5-1

2. Use the rules of Boolean Algebra to reduce the expression written in problem 1.

3. Draw the new logic diagram for the reduced expression in problem 2.

4. Draw a logic diagram for the following Boolean Expression.

$$Y = AB + C\overline{B} + D$$

5. Draw a logic diagram for the following Boolean Expression.

$$X = \overline{AB + \overline{C}B}$$

6. Write a Boolean Expression for the logic diagram shown in Figure 5-2.

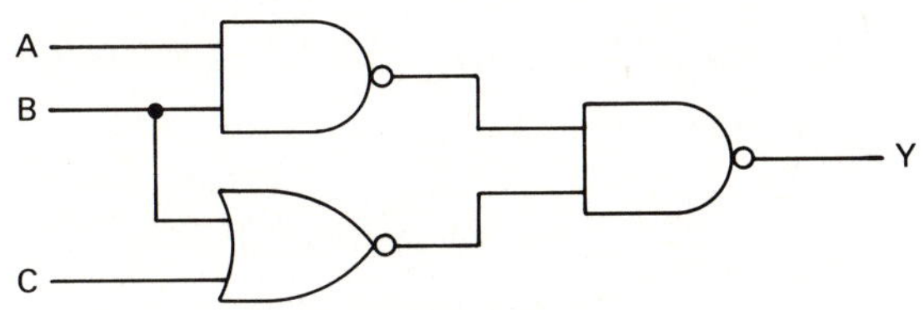

Figure 5-2

7. Use the rules of Boolean Algebra to reduce the Boolean Expression in problem 6.

8. Draw a logic diagram that illustrates the distributive law of multiplication.

9. Simplify the following Boolean Expression using the rules of Boolean Algebra.

$$Y = C[\overline{A}(B + \overline{C}) + A\overline{B}]$$

10. Use DeMorgan's theorem to reduce the following Boolean Expression.

$$X = \overline{AB} + CB$$

11. Change the logic diagram in Figure 5-3 to all NAND gates using the technique of bubble pushing.

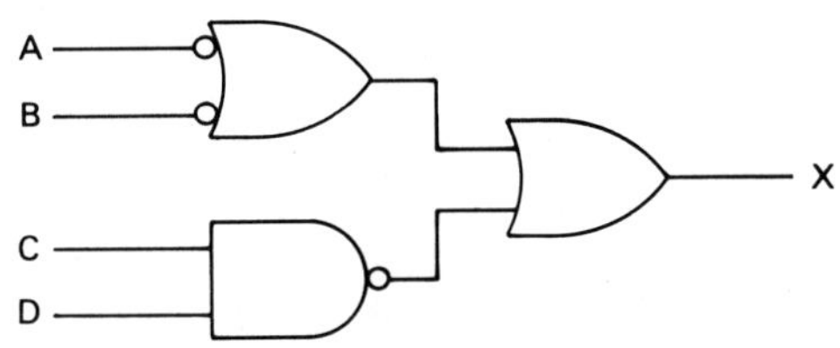

Figure 5-3

12. Write the Boolean Expression for the circuit of Figure 5-4 using DeMorgan's theorem convert the circuit to AND, OR and Inverter logic.

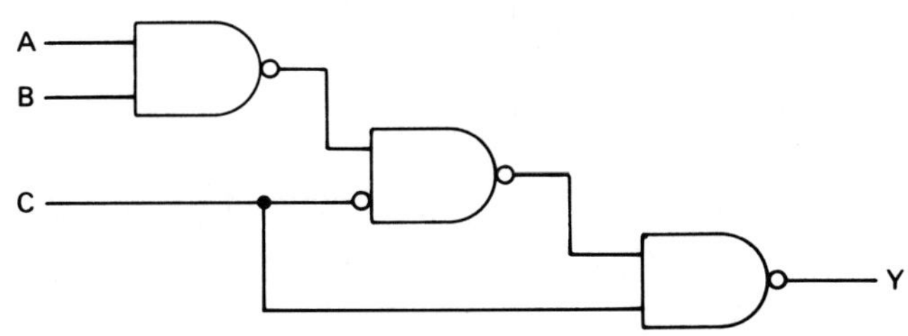

Figure 5-4

13. Develop a truth table for the following Boolean Expression.

$$\overline{A} + \overline{B} \bullet \overline{C} = X$$

14. Write a SOP Boolean Expression from the following truth table.

A	B	C	Y
0	0	0	1
0	0	1	0
0	1	0	1
0	1	1	0
1	0	0	1
1	0	1	1
1	1	0	1
1	1	1	0

15. Use a k-map to reduce the Boolean Expression of problem 14.

16. Show how to construct the NOR logic function using only NAND logic gates.

17. Show how the AND logic function can be accomplished using only NOR logic gates.

18. Reduce the following Boolean Expression using a k-map.

$$Y = \overline{A}\,\overline{B}\,\overline{C}\,\overline{D} + \overline{A}BCD + ABCD + \overline{A}\,\overline{B}\,\overline{C}D + \overline{A}BC\overline{D}$$

19. Use a 74LS54 AOI gate to implement the following Boolean Expression.

$$Z = \overline{\overline{A}B(A + B + C)} + \overline{A}CD$$

20. Reduce the following Boolean Expression using a k-map.

$$W = AB + \overline{A}BC + \overline{ABC} + A\overline{BC}$$

6

CHAPTER

Exclusive-OR and Exclusive-NOR Gates

The Exclusive-OR Gate

- The Exclusive-OR output is true if either input is true but not both.
- The Exclusive-OR function can be expressed in Boolean Algebra as

$$X = \overline{A}B + A\overline{B}.$$

- The Exclusive-OR function is used so often that there is a special Boolean symbol for it. $X = A \oplus B$ (read X equals A exclusive-OR B)
- The Exclusive-OR function also has its own logic symbol.

The Exclusive-NOR Gate

- The Exclusive-NOR is the complement of the Exclusive-OR gate.
- The Exclusive-NOR gate is also called an "equality gate" since both inputs must be equal to get a high out.
- The Exclusive-NOR symbol is the same as the Exclusive-OR symbol with a bubble on the output (for the inversion).
- The Boolean Expression for the Exclusive-NOR is

$$X = AB + \overline{A}\overline{B}$$

- Both the Exclusive-OR and the Exclusive-NOR are two input gates only.

Parity Generator/Checker

- One important application of the use of an Exclusive-OR gate is to generate parity.
- Parity is used to detect errors in transmitted data caused by noise or other disturbances.
- A parity bit is an extra bit that is added to a data word and can be either odd or even parity.
- In an even parity system, the sum of all the bits (including the parity bit) is an even number.

  In an odd parity system the sum of all the bits must be an odd number.
- The circuit that creates the parity bit at the transmitter is called the parity generator.

  The circuit that determines if the received data is correct is the parity checker.
- Parity is good for detecting a single bit error only.
- The parity generator and the parity checker can both be built using Exclusive-OR gates.
- To generate even parity the bits of data are Exclusive-ORed together in groups of two until there is only a single output left. This output is the parity bit.
- To generate odd parity, simply invert the even parity.

  The last gate can be an Exclusive-NOR gate.
- To check parity first a new parity bit must be generated over the data that was received. This parity bit is then compared to the received parity bit. If they are the same, then all is ok. The comparison can be done with an Exclusive-OR gate.
- The 74280 is a TTL IC parity generator/checker circuit.

System Design Applications

- Exclusive-OR circuits can be used to generate a parity check bit.
- Exclusive-NOR circuits can be used to compare two binary data words and determine if they are exactly the same.
- Exclusive-OR gates can be used to complement (controlled inverter) a binary data word.

Additional Notes

The Exclusive-OR function is one of the most useful of the special logic functions. The Exclusive-OR is sometimes also called the "Either/OR but not Both" function. With only two inputs, the output is true if either of the inputs is true but not both of the inputs. Another way to say this is that the output is true if both of the inputs are the opposite of each other.

Although the Exclusive-OR is a special function, it is most convenient to use the Exclusive-OR in its SOP form ($X = \overline{A}B + A\overline{B}$). Although the special symbol $\oplus$ can be used to identify the Exclusive-OR, there are no rules in the Boolean Algebra to deal with this special formula. If the Exclusive-OR symbol is encountered in a Boolean Expression, it should be converted to its SOP form before any reduction process is started.

TEST 6

1. Complete the truth table for an Exclusive-OR gate.

2. Draw an Exclusive-OR function using AND, OR and Inverter gates.

3. Draw the logic symbol for an Exclusive-NOR gate.

4. Use Exclusive-OR gates to generate even parity over a five-bit data word.

5. Show from the truth table how an Exclusive-OR gate can be used to invert the data on one input if the other input is a special control function.

6. Determine odd parity for each of the following data words:

1011101 11110111 1001101

7. Show the circuit that uses two Exclusive-NOR gates to compare two two-bit data words.

8. How is odd parity generated differently from even parity?

9. What is the difference between a parity generator and a parity checker?

10. Why is an Exclusive-NOR gate also called an equality gate?

7
CHAPTER

Arithmetic Operations and Circuits

Binary Arithmetic

Addition

- There are four basic rules to adding two binary digits.

  0 + 0 = 0 carry 0

  0 + 1 = 1 carry 0

  1 + 0 = 1 carry 0

  1 + 1 = 0 carry 1

- Binary digits are added two at a time and any carry must be carried over to the next higher column of digits.
- To get the sum of three digits, add the first two and then add the sum to the third digit.
- To add large binary numbers add one column of digits starting with the least significant position. Add any carry into the next significant column.

Subtraction

- There are four possible combinations when subtracting two binary digits.

  0 - 0 = 0 borrow 0

  0 - 1 = 1 borrow 1

  1 - 0 = 1 borrow 0

  1 - 1 = 0 borrow 0

- The borrow from a column of digits must be subtracted from the next most significant column.
- To subtract a large binary number from another large binary number, a borrow may need to be carried over several bit positions.

Multiplication

- If the bit of the multiplier is a 1 then write down the multiplicand starting with the LSB under the bit of the multiplier.
- If the bit of the multiplier is a 0, write a 0 under the bit of the multiplier.
- Repeat this procedure for each bit of the multiplier, then add all these products to get the final result.

Division

- Binary division follows the same procedure as decimal division except the numbers are binary.
- Any number left over after the last subtraction would be a remainder.

Two's-Complement Representation

- The two's-complement method is one of the most popular methods of representing binary numbers in digital circuits.
- Most computers use an 8-bit or a 16-bit binary word.
- The MSB is the sign bit and a 0 means positive and a 1 means negative.
- Positive numbers are in true form and negative numbers are in two's-complement form.
- The range of positive numbers is 0 to (2^{N-1} - 1) where N is the number of bits. (0 to 127 for 8-bit numbers)
- The range of negative numbers is -1 to -2^{N-1} (–1 to -128 for 8-bit numbers)
- To find the two's-complement of a number:
  1. complement each bit of the word (1's complement).
  2. add 1 to the number.

Two's-Complement Arithmetic

- In order to add signed numbers, simply perform the standard addition. If the numbers are correctly signed, the answer will come out with the correct sign.
- In order to subtract signed numbers, take the two's-complement of the number being subtracted and add the two numbers.
- When adding and subtracting be sure the answer does not exceed the range of the signed numbers.

  (+127 to -128 for 8-bit numbers; +32,767 to -32,768 for 16-bit numbers)

Hexadecimal Arithmetic

- Hexadecimal is a base 16 number system.
- To add hexadecimal numbers:
  1. Add the two hex digits by working with their decimal equivalents.
  2. If the decimal sum is less than 16 write down the hex equivalent.
  3. If the decimal sum is 16 or more, subtract 16, write down the hex result and carry 1 to the next-more-significant column.
- Hexadecimal addition can also be accomplished by converting the hexadecimal numbers to binary and adding the binary numbers.
- Subtraction of hexadecimal numbers:
  1. Subtract the decimal equivalents of the hex digits.
  2. If a borrow is needed, subtract 1 from the more-significant digit and add a decimal 16 to the digit being borrowed to.
  3. Write down the hex equivalent of the decimal difference.
- Hexadecimal subtraction can be accomplished by converting the hexadecimal number to binary and then performing binary subtraction.

BCD Arithmetic

- BCD is a binary code of the ten decimal digits. It is not a binary equivalent.
- To perform BCD addition
  1. Add the BCD digits as regular binary numbers.
  2. If the sum is 9 or less and no carry was generated, it is a valid BCD digit.
  3. If the sum produces a carry, the sum is invalid and the number 6 (0110) must be added to the digit.
  4. If the sum is greater than nine, the sum is invalid and the number 6 (0110) must be added to the digit.
  5. Repeat for each of the BCD digits.

Arithmetic Circuits

- The basic process of adding two binary digits can be accomplished with a half-adder circuit.
- In a half-adder, the sum output is the Exclusive-OR of the two input bits. The carry output is the logical AND of the two input bits.
- If a third input is required (carry in from previous-significant digit) then a full-adder circuit must be used.
- A full-adder is really two half-adders with the carry outputs ORed together.
- The full-adder and the half-adder can be represented with a block symbol. The half-adder has two inputs and the full-adder has three inputs.
- In order to add two binary words with several bits, an adder circuit is required for each bit of the data word. To make the circuit complete, the carry input for each full-adder is connected to the carry output of the previous least-significant bit.

Four Bit Full-Adder ICS

- There are a number of ICs available that contain adder circuits.
- The 7483 and 4008 are examples of ICs with 4 full-adder circuits.
- The 7483 and 4008 use a fast-look-ahead carry circuit to speed up the circuit propagation time. Without this circuit the carry would have to ripple through all 4 bits.

System Design Applications

- Exclusive-OR gates can be used to perform a controlled inversion of a data word.
- The controlled inversion combined with an additional 1 on the carry input of the LSB of a set of full-adders will give the two's-complement of a number.
- Subtraction can be accomplished by taking the two's-complement of a number and adding.
- BCD addition can be accomplished also with full-adders.

  One set of adder circuits is used to do the original addition. Some addition logic is necessary to detect a carry or an illegal number. A second set of adder circuits is then used to add six to the first sum.
- The 74HC583 is a single chip that will perform all the necessary steps in order to add 2 BCD digits.

Arithmetic/Logic Units

- An ALU can perform several different arithmetic or logical operations.
- The 74181 is a 4 bit ALU with 16 logic and 16 arithmetic functions.

Additional Notes

Signed numbers can be represented with the two's-complement form. In this form the most significant bit is the sign bit. With one bit used as a sign then the magnitude of the number is reduced by one bit. For a data word with 8-bits, the normal capacity would be 256 numbers. With the sign bit there are 128 positive numbers (0 to +127) and 128 negative numbers (-1 to -128). This is still 256 numerical combinations.

Care must be taken, however, when working with signed numbers not to exceed the range of positive or negative numbers. For example: if the number +2 (0000 0010) is added to +126 (0111 1110) the results will be 1000 0000.

This is not +128 but a negative number -127. Two positive numbers added together resulted in a negative answer. This is a condition called overflow. Most computers have special circuits to detect an overflow condition.

TEST 7

1. Add the following binary numbers.

0010 0110	0011 1011	0011 1100
+0101 0101	+0001 1110	+0001 1111

2. Subtract the following binary numbers.

0101 1000	1010 0011	1101 1110
-0010 0011	-0011 1000	-0101 0111

3. Multiply the following binary numbers.

1010	1011	1001
× 0011	× 0111	× 1010

4. Divide the following binary numbers.

$0110\overline{)001100}$ $\quad$ $0011\overline{)000110}$ $\quad$ $0010\overline{)001000}$

5. Determine the two's-complement of each binary number.

00110 $\quad$ 00011 $\quad$ 11101

6. Convert each of the signed decimal numbers to an 8-bit signed binary number (two's-complement).

+7   -3   -12

7. Convert each of the following signed binary numbers (two's-complement) to a signed decimal number.

00000101   11111100   11111000

8. Convert each of the decimal numbers to two's-complement form and perform the addition in binary.

+13	-10
add -7	add +15

9. Perform subtraction on each of the following binary numbers by taking the two's-complement of the number being subtracted and then adding it to the first number.

01001	01100
00011	00111

10. Convert each of the decimal numbers to 8-bit two's-complement form and then perform subtraction by taking the two's-complement and adding.

-18	+30	-20
sub +4	sub +17	sub -10

11. Add the following hexadecimal numbers.

3C	14	3B
+25	+28	+DC

12. Subtract the following hexadecimal numbers.

47	34	FA
-25	-1C	-2F

13. Add the following BCD numbers.

0110	0111	1001
0101	1000	1000

14. What logic function is the sum output of a half-adder?

15. Sketch a half-adder circuit using only AND, OR and Inverter gates.

16. Could the sum output of a full adder be used as a two-bit parity generator?

17. What is the difference between a full-adder and a half-adder?

18. Sketch a block diagram of a circuit to add two 3-bit binary numbers.

19. Why is a fast-look-ahead carry circuit used in the 7483 4-bit full adder?

20. Explain what an ALU circuit is and how it differs from a full adder.

8

CHAPTER

Code Converters, Multiplexers, and Demultiplexers

Comparitors

- A comparitor will evaluate two binary strings and output a 1 if the two strings are exactly the same.
- The Exclusive-NOR (equality gate) is used to perform the comparison.
- One Exclusive-NOR is used per pair of binary bits and the outputs of all Exclusive-NORS are ANDed together.
- The 7485 is a 4-bit magnitude comparitor.

  A magnitude comparitor will determine if A = B, A > B or A < B.
- Expansion inputs are provided on the 7485 so that word sizes larger than 4-bits may be compared.

Decoders

- A decoder circuit converts a binary code on the input to a single output representing the numeric value of the code.
- A binary-to-octal decoder converts 3 binary bits into a 1-of-8 output.
- Decoders can be designed with either active high outputs or active low outputs.
- A BCD decoder converts a 4-bit BCD code on the input to a 1-of-10 output.
- A hexadecimal decoder converts a 4-bit binary code on the input to a 1-of-16 output.

- Any decoder that uses bubbles on the outputs has an active low output. No bubbles mean that the outputs are active high.
- Many decoder chips have a separate enable input to turn the chip on and off.
- The 74138 IC is an octal decoder with three enable inputs and active-low outputs.
- The 7442 IC is a BCD decoder with active-low outputs.
- The 74154 is a hex decoder with two active-low enable inputs and active-low outputs.

Encoding

- An encoder circuit will convert a single input out of several to a binary code on the output.
- A decimal-to-BCD encoder will convert one active input out of ten to a BCD code on the output.
- An octal-to-binary encoder will convert one-of-eight inputs to a 3-bit binary code on the output.
- A priority encoder such as the 74147 will output the proper code for the higher value of two or more active inputs and provide decimal-to-BCD encoding.
- The 74148 IC is an octal-to-binary priority encoder.

Code Converters

- A code converter circuit will convert coded information in one form to a different coding form.
- One example of code conversion is to convert BCD to straight binary.
- The weighting of BCD bits is not the same as straight binary. The weight of each bit within a BCD digit is 1, 2, 4, 8 respectively; but, each digit must be multiplied by its decimal positional value (1, 10, 100, 1000, etc.).
- The 74184 IC will convert 1 1/2 BCD digits to straight binary. (the LSB is the same in BCD and in binary).
- The 74158 IC is used to convert 6 binary bits into a BCD code (again the LSB is feed around the chip).
- Converters are also available that convert BCD input to a seven segment code output such as the 7447 IC.
- The Gray code is another useful code that is primarily used to indicate angular position.
- The Gray code differs from regular binary in that between adjacent counts only 1 bit changes at a time.
- Conversion from binary to Gray and from Gray to binary can be done with the Exclusive-OR gate.

Multiplexers

- The multiplexer (also known as a data selector) will select data from several transmission lines to be gated to the output transmission line.
- The multiplexer will also have a number of control inputs that are used to select the appropriate data channel for input.
- The number of data inputs is equal to 2^N where N is the number of control select leads.
- The 74151 is a eight-line multiplexer with three select lines.
- An additional input on the 74151 is an enable input which can be used to expand the size of the multiplexer.
- A multiplexer can also be used to implement a SOP Boolean Expression in combinational logic.

Demultiplexers

- A demultiplexer (data distributor) will select a specific output data transmission path from many for the input data to be sent to.
- The demultiplexer has several control select lines which are used to determine the selected output transmission path.
- The number of data outputs is equal to 2^N where N is the number of control select leads.
- The 74139 is a 1-of-4 decoder that can also be used as a demultiplexer.
- To use the 74139 as a demultiplexer, connect the incoming data to the enable input and use the A0 and A1 inputs as the control select lines.
- The 74154 1-of-16 decoder can also be used as a 16 line demultiplexer.
- Several CMOS chips use a transmission gate as part of the multiplexer/demultiplexer. This means that the gate is bidirectional and can also handle AC signals as well as binary 1's and 0's.

System Design Applications

- Decoders are often used in microprocessor systems to decode the address information from the microprocessor in order to select the correct memory chip. A popular IC for this application is the 74138 or 74HCT138.
- Encoders are often used in microprocessor or in microcontroller applications to combine or group a large number of inputs down to a smaller binary count of the input activity. The 74148 is a popular IC encoder chip for these kinds of applications.
- Microprocessor based systems with serial input and output ports can use multiplexers and demultiplexers to route serial data to and from a number of different destinations. Multiplexers are used to combine several inputs into one and demultiplexers, or data distributors are used to send a single output to one of several destinations.
- An analog multiplexer can be used as a staircase generator which in turn can be used to provide a multiple input for an oscilloscope.

- A multiplexer/demultiplexer can be used to multiplex a 7-segment display system.

Additional Notes

Multiplexers and demultiplexers have many applications. The process of data selection and data routing have already been shown. Another important application is in parallel-to-serial conversion and in serial-to-parallel conversion. A multiplexer can be used to convert parallel data to serial data. The parallel data is applied to the data inputs and a binary count sequencer is connected to the control inputs. This means that the inputs will appear on the output one-at-a-time and in numerical sequence. Thus the parallel data will be applied to the output serially. The inverse (serial-to-parallel) can be accomplished with a demultiplexer.

TEST 8

1. How many Exclusive-NOR gates would be required for an 8-bit comparitor circuit?

2. Explain the difference between a comparitor circuit and a magnitude comparitor circuit.

3. What do the mathematical symbols A<B and A>B mean?

4. For the following conditions on a 7485 magnitude comparitor, what will be the state of each of the three outputs?

A0 = 0	B0 = 1	IA < B = 0
A1 = 1	B1 = 1	IA = B = 1
A2 = 1	B2 = 0	IA > B = 0
A3 = 0	B3 = 0	

5. How many inputs are required for a 1-of-16 decoder?

6. How many outputs are on a BCD decoder?

7. How can the active condition (high or low) or the decoder output be determined from the logic symbol?

8. What is the purpose of an enable input to a decoder?

9. How many possible outputs would a decoder have with a 6-bit binary input?

10. How is an encoder different from a decoder?

11. How many inputs will a decimal-to-BCD encoder have?

12. If two inputs are active on a priority encoder, which will be coded on the output?

13. From the following list of input conditions, determine the state of the 5 output leads on a 74148 octal-to-binary encoder.

I0 = 1	I3 = 1	I6 = 1
I1 = 1	I4 = 0	I7 = 1
I2 = 1	I5 = 1	EI = 0

14. Use the weighting factors to convert the following BCD numbers to binary.

0101 0011   0010 0110 1000

15. How many 74184 BCD-to-binary converters would be required to convert two complete BCD digits to a binary number?

16. Why is the Gray code more practical to use when coding the position of a rotating shaft?

17. Show the logic diagram for converting a 6-bit Gray code into a binary number.

18. How many select lines would be required for an 8-line to 1-line multiplexer?

19. What is the function of an enable input on a multiplexer chip?

20. What is the status of the inputs S0, S1 and S2 of the 74151 eight-line multiplexer in order for the output Y to be a copy of input I5?

21. Show how to use a 74151 eight-line multiplexer to provide the logic for the following Boolean Expression.

$$Y = \overline{A}BC + A\overline{B}C + AB\overline{C} + ABC$$

22. Why is a demultiplexer called a data distributor?

23. Show how a 1-of-8 decoder can be used as an 8-line demultiplexer.

24. Why can a CMOS IC be used as both a multiplexer and a demultiplexer?

9

CHAPTER

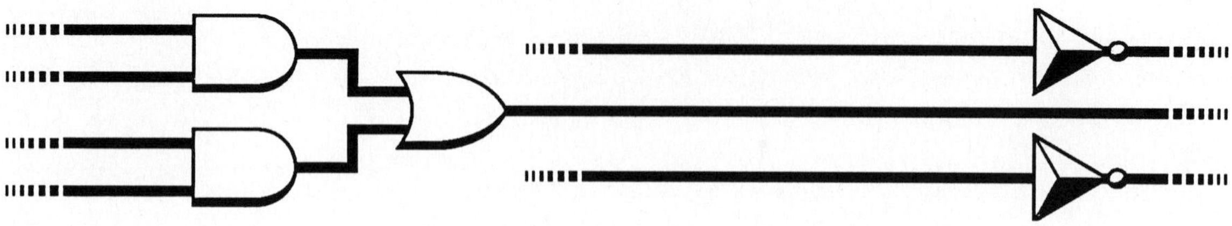

Logic Families and their Characteristics

The TTL Family

- The input of a TTL logic gate is characterized by a multi-emitter transistor.
- Each emitter is one input to the logic gate. A two input gate will have two emitters and a four input gate will have four emitters.
- The output stage of most TTL logic is a totem-pole circuit.
- When at least one input to a TTL NAND gate is low, the bottom transistor of the totem-pole output (Q4) is OFF and the top transistor (Q3) is ON, making the output a logic high.
- When all inputs to a TTL NAND gate are high, the top transistor of the totem-pole output (Q3) is OFF and the bottom transistor (Q4) is ON, pulling the output low.

TTL Voltage and Current Ratings

- The input currents that a gate load draws:

I_{IH} - (Input High Current) - 74XX = 40 uA

I_{IL} - (Input Low Current) - 74XX = -1.6 mA

- The output currents that a driving gate can supply:

I_{OH} - (Output High - source current) - 74XX = -400 uA

I_{OL} - (Output Low - sink current) - 74XX = 16 mA

- The minus sign indicates current leaving the gate.
- Fan-out is the number of gate inputs that can be driven by a single output of the same family.

$$\text{Fan-out} = I_{OH}/I_{IH} \text{ and Fan-out} = I_{OL}/I_{IL}$$

(The smaller of these two values would be used if they are not the same value.)

- The actual output current from a gate would be the number of gate inputs connected to the circuit multiplied by the input current of each gate.
- The voltage on the output of a TTL gate:

V_{OH} - (output high) - 74XX = 2.4 V(min), 3.4 V(typ)

V_{OL} - (output low) - 74XX = 0.2 V(typ), 0.4 V(max)

- The required limit of the input voltage in order to guarantee operation:

V_{IH} - (input high) - 74XX = 2.0 V(min)

V_{IL} - (input low) - 74XX = 0.8 V(max)

- The noise margin is the difference between the guaranteed output voltage level and the required input voltage level of a logic gate.

low level noise margin = $V_{IL} - V_{OL}$ = 0.4 V(74XX)

high level noise margin = $V_{OH} - V_{IH}$ = 0.4 V(74XX)

Other TTL Considerations

- For a non-ideal waveform the rise time is a measure of the time required for the leading edge of the waveform to rise from 10% to 90% of its nominal amplitude.
- For a non-ideal waveform the fall time is a measure of the time required for the trailing edge of the waveform to fall from 90% to 10% of its nominal amplitude.
- The propagation delay is a measure of the time it takes the output of a logic gate to respond to a change on the input. The propagation delay is caused by the circuit of the logic gate.
- t_{PLH} is the propagation delay as the output waveform moves from low-to-high.
- t_{PHL} is the propagation delay as the output waveform moves from high-to-low.
- The current drawn from the power supply by an IC is called I_{CC}. This is usually given as two values:

  I_{CCH} is supply current when all outputs are high.

  I_{CCL} is supply current when all outputs are low.
- The average supply current is the sum of I_{CCH} and I_{CCL} divided by two.
- The power dissipation (P_D) of an IC is the supply voltage Vcc multiplied by the average supply current Icc(avg).
- When the upper transistor of a TTL totem-pole output is removed, the circuit is called an open collector output.

- Any open collector output requires an external pull-up resistor for the high level current path.
- Special open collector circuits called buffer/drivers are available for applications requiring large sink currents.
- The open collector circuit is used whenever two or more outputs of logic gates are to be connected together.
- The connecting together of two or more outputs forms a logical AND function and is called a wired-AND logic.
- Many of the standard TTL logic gates are also available in open collector arrangements.
- Unused inputs on AND and NAND gates should be connected high while unused inputs on OR and NOR gates should be tied low.
- Any unused gate on a chip should have the inputs connected so that its output is high. (AND and OR gate inputs high; NAND and NOR gate inputs low.)
- Decoupling capacitors (.01 uF to 0.1 uF) should be used on all TTL IC's between the Vcc and GND pin. The capacitors help to reduce TTL switching noise.

Improved TTL Series

- Schottky TTL (74SXX) uses schottky-clamped transistors to reduce the propagation delay by 4 and increase the power consumption by only 2.
- Low-power schottky (74LSXX) has reduced power consumption from the schottky TTL.
- Advanced low-power schottky (74ALSXX) has further reductions in propagation delay and power consumption.
- Fast TTL (74FXX) has reduced propagation delay from the LS and ALS series.

The CMOS Family

- The CMOS family uses a complementary pair (one N-channel and one P-channel) of MOSFET devices.
- The CMOS family has a high input impedance and a very low power consumption.
- If the input to a CMOS gate is high, the N-channel device will be ON pulling the output to Ground.
- If the input to a CMOS gate is low, the P-channel device will be ON raising the output to V_{DD}.
- CMOS logic is available in either NAND or NOR configurations.
- CMOS devices must be handled very carefully to avoid damage from static discharge.
- CMOS chips are available in almost all the same logic configurations as TTL as well as several that are unique to CMOS.

- There are several families of CMOS logic:
    4000 series - the original series.
    40H00 series - faster.
    74C00 series - pin compatible to TTL
    74HC00 series - high speed pin compatible to TTL
    74HCT00 series - input/output voltage compatible to TTL
- The latest series of logic to come out is the BiCMOS series which combines the best features from both the bipolar transistors and CMOS transistors.
- These BiCMOS combine high speed switching with very low power consumption. They also have a very low power consumption when idle (or inactive).
- The BiCMOS logic is usually limited to microprocessor bus interface logic.

Emitter-Coupled Logic

- ECL logic is very fast but has an increased power consumption.
- ECL uses a differential amplifier as its basic circuit.
- ECL is very fast because the transistors are not allowed to saturate.
- The voltage levels for ECL (-0.8 V and -1.7 V) are unusual and not compatible to TTL or CMOS logic.
- The ECL logic gates usually have both an OR and a NOR output.
- New technologies are continually being developed in an effort to improve the logic.

Comparing Logic Families

- Major parameters for comparing logic families are the propagation delay, the power dissipation and the speed-power product.
- The power dissipation of CMOS logic depends heavily upon the operating frequency of the logic.

Interfacing Logic Families

- When using TTL logic to drive CMOS logic the high level voltage of the TTL output must be increased with a pull-up resistor.
- When CMOS logic is expected to drive TTL logic, the CMOS gate may not have the current capability to drive the TTL gate. In this case, special buffer drivers (4050B) gates can be used.
- Any time one logic family is connected to another logic family, the input and output voltage and current requirements for each family should be reviewed to determine the interface needs for the circuit connection.
- When the two logic families being interfaced have different power supplies then special circuits called level shifters or translators must be used.

Additional Notes

Another very useful function that is available with TTL logic is the tri-state buffer. The tri-state buffer is a standard TTL logic gate to which has been added an additional enable input. This enable input, when enabled, allows the buffer to react as a standard TTL inverter. When the enable input is disabled, however, the buffer has a third condition on the output. This is called the high-impedance state (high-Z) and is characterized by no current (sink or source) in the output lead. The enable input (when not active) will allow both transistors of the totem-pole output to be turned OFF. This means that no current can flow in the output circuit and it is as if the output pin is disconnected from the circuit.

The tri-state buffer is useful when interfacing TTL logic to a common connection (wired-AND). If the open collector gate is used for the common connection, the overall speed is reduced because the open collector circuit has a larger propagation delay than standard TTL. The tri-state buffer, on the other hand, retains the same propagation delay as the rest of its family. Therefore, the tri-state buffer operates faster than the open collector circuit. In order to use the tri-state buffer, however, it must be guaranteed that only one buffer will be enabled at a time.

TEST 9

1. Using the schematic diagram of a TTL NAND gate, determine the state of each transistor (ON or OFF) when all inputs are high.

2. If all inputs to a TTL NAND gate are low, what is the ON, OFF condition of each transistor in the circuit?

3. What is the difference between the 54XX and 74XX series of TTL logic gates?

4. From the following specifications determine the fan-out for the logic family.

$I_{OH} = 20uA$ $\quad$ $I_{IH} = 2.5uA$

$I_{OL} = 32ma$ $\quad$ $I_{IL} = 2mA$

5. Using the above specifications, determine the amount of current in the output of a logic gate if the output of the gate is a logic low and connected to 4 inputs.

6. Why is the maximum value of V_{OL} and the minimum value of V_{OH} used to determine the noise margin rather than the typical value for these parameters?

7. Determine the rise time and the fall time of the waveform shown in Figure 9-1.

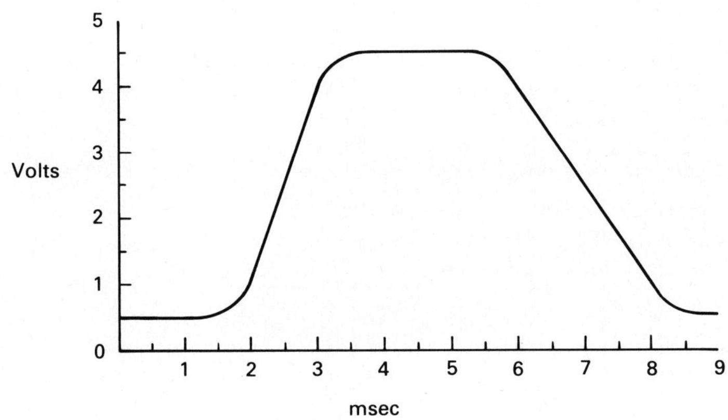

Figure 9-1

8. From the waveforms shown in Figure 9-2, determine the propagation delay.

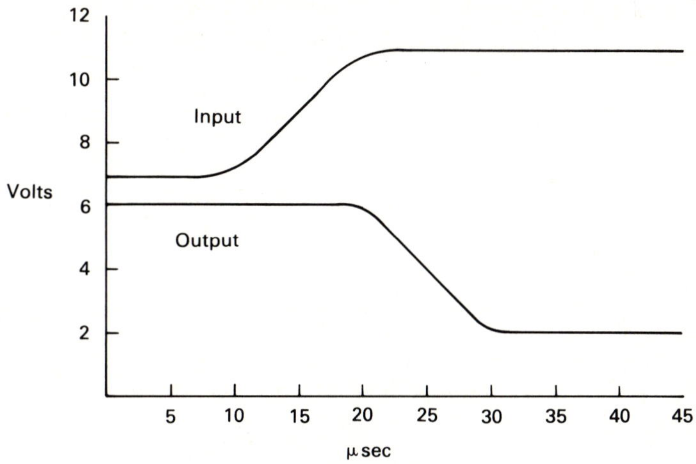

Figure 9-2

9. Would the propagation delay in Figure 9-2 be labeled t_{PLH} or t_{PHL}?

10. Assume that a particular IC has a supply voltage (Vcc) equal to +5V and I_{CCH} = 10 mA and I_{CCL} = 23 mA, what is the power dissipation for the chip?

11. Why is a pull-up resistor needed for an open collector gate?

12. What should be done with unused inputs to a TTL NAND gate?

13. Why is a decoupling capacitor needed for TTL IC's and where should it be connected?

14. What is the advantage of using low-power schottky (LS) over standard TTL logic?

15. Which logic family is characterized by a multi-emitter transistor on the input?

16. Why must CMOS devices be handled with care?

17. How does the 4000 series of CMOS logic compare in terms of speed and power dissipation to the standard family of TTL logic?

18. What is the difference between the 74HC00 series and the 74HCT00 series of CMOS logic?

19. What is the major advantage to ECL logic?

20. How can ECL have both a NOR and OR output?

21. How is the speed-power product of a logic family determined?

22. Why is a pull-up resistor needed when connecting TTL logic to CMOS logic?

23. Can a 74HCMOS logic gate directly connect to a 74ALSTTL gate?

24. When is a level shifter circuit needed in interfacing logic?

25. How many 74LSTTL logic gates can be driven from a 74TTL gate?

10

CHAPTER

Flip-Flops and Registers

S-R Flip-Flop

- A set-reset flip-flop can be constructed from cross-coupled NOR gates.
- The set and reset inputs of the cross-coupled NOR flip-flop are high active.
- A set input will force the Q output high and the $\overline{Q}$ output low.
- A reset input will force the Q output low and the $\overline{Q}$ output high.
- With neither the set input nor the reset input active, the flip-flop is in a "hold" condition and will "remember" the last input condition.
- A S-R flip-flop can also be made using cross-coupled NAND gates.
- The cross-coupled NAND flip-flop has low active set and reset inputs. These can be made high active by using inverters.
- A flip-flop is "latched" when the Q output holds the last input condition.
- A flip-flop is said to be transparent when the Q output responds immediately to a change on the input.
- A register is a group of flip-flops used to store a binary word. One flip-flop is needed for each bit in the data word.
- AND gates can be used to "strobe" or enable data gated into a register.

Gated S-R Flip-Flop

- A gate input is added to the S-R flip-flop to make the flip-flop synchronous.
- In order for the set and reset inputs to change the flip-flop, the gate input must be active (high).
- When the gate input is low, the flip-flop remains in the hold condition.

Gated D Flip-Flop

- The D (data) flip-flop has a single input that is used to set and to reset the flip-flop.
- When the gate is high, the Q output will follow the D input.
- When the gate is low, the flip-flop is latched.

The Integrated-Circuit D Latch (7475)

- The 7475 contains 4 transparent D latches with a common enable (gate) on latches 0 and 1 and another common enable on latches 2 and 3.
- When Q follows D (latch enabled) the latch is said to be transparent.

The Integrated-Circuit D Flip-Flop (7474)

- The 7474 is an edge-triggered device. The Q output will change only on the edge of the input trigger pulse.
- The small triangle on the clock (Cp) input of the symbol indicates that the device is positive edge-triggered.
- The D and the clock inputs are synchronous inputs.
- The set (S_D) and reset (R_D) inputs are asynchronous. They operate independent of D and Cp.
- The bubbles on the set and reset inputs indicate that they are low active.

Master-Slave J-K Flip-Flop

- The J-K flip-flop has a toggle mode of operation when both J and K inputs are high. Toggle means that the Q output will change states on each active clock edge.
- J, K and Cp are all synchronous inputs.
- The master-slave flip-flop is constructed with two latches. The master latch is loaded with the condition of the J-K inputs while the clock is high. When the clock goes low, the slave takes on the state of the master and the master is latched.
- The master-slave is a level-triggered device.
- The master-slave can interpret unwanted signals on the J-K inputs.

The Edge-Triggered J-K Flip-Flop

- The edge-triggered J-K will only accept the J and K inputs during the active edge of the clock.

- The small triangle on the clock input indicates that the device is edge-triggered.
- A bubble on the clock input indicates that the device responds to the negative edge. No bubble would indicate a positive edge-triggered device.

The Integrated-Circuit J-K Flip-Flop (7476, 74LS76)

- The 7476 is a master-slave J-K and the 74LS76 is a negative edge-triggered J-K flip-flop.
- Both chips have the same pin configuration.
- Both chips have synchronous inputs of J, K and Cp.
- Both chips have asynchronous inputs $\overline{S}_D$ and $\overline{R}_D$.
- The J-K flip-flop can be made into a D flip-flop by bringing the data input into the J and the inverse of the data input into the K input.

An Octal D Flip-Flop in a Microcontroller Application

- The 74HCT273 is an 8-bit D flip-flop register.
- All 8 flip-flops are controlled by a common clock.
- The 273 can be used as an update and hold register.

Additional Notes

The master-slave flip-flop is an earlier development than the edge-triggered flip-flop. The master-slave had a problem with the flip-flop picking up unwanted conditions on the J-K inputs while the master was enabled (clock high). That is why the 7476 (being the earlier chip) is a master-slave and the 74LS76 is an edge-triggered flip-flop. The edge-triggered flip-flop will not have this problem because the flip-flop can only change one time during the clock cycle. The only time the flip-flop can change is during the active transition of the clock.

Another modification that has helped the master-slave flip-flop is a feature called data lockout. With the data lockout feature, the master latch is prevented from changing 20 to 30 usec. after the clock goes high. This, in effect, allows the master to pick up the J-K inputs only during the leading edge of the clock and then the master is transferred to the slave during the negative edge of the clock. For the most part, though, the edge-triggered device is an improvement over the master-slave flip-flop.

TEST 10

1. Does the cross coupled NOR flip-flop have active high or active low set and reset inputs?

2. How can the cross coupled NAND flip-flop be made to have active high S-R inputs?

3. What is the hold condition of a flip-flop?

4. Show the pin connections to wire a 7400 IC as a S-R flip-flop with active high inputs.

5. Sketch the Q output waveform for the circuit of Figure 10-1.

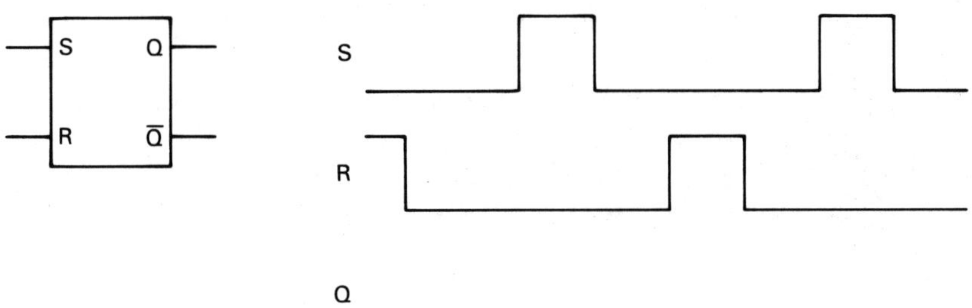

Figure 10-1

6. Why are the S and R inputs of a gated flip-flop said to be synchronous?

7. Sketch the Q output for the circuit shown in Figure 10-2.

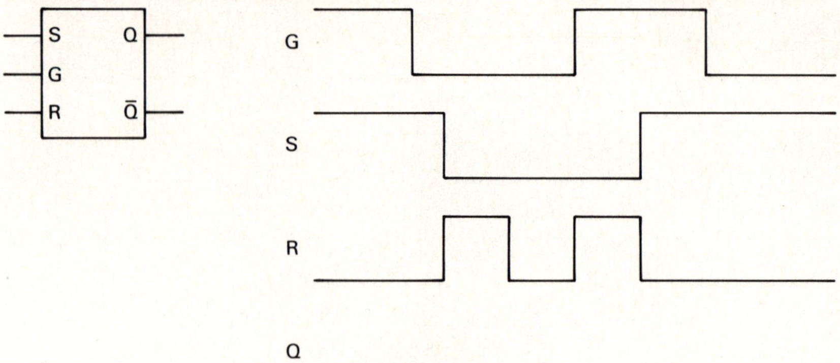

Figure 10-2

8. When is a flip-flop said to be transparent?

9. Explain the difference between a gated S-R flip-flop and a gated D flip-flop.

10. Sketch the output waveform for the circuit shown in Figure 10-3.

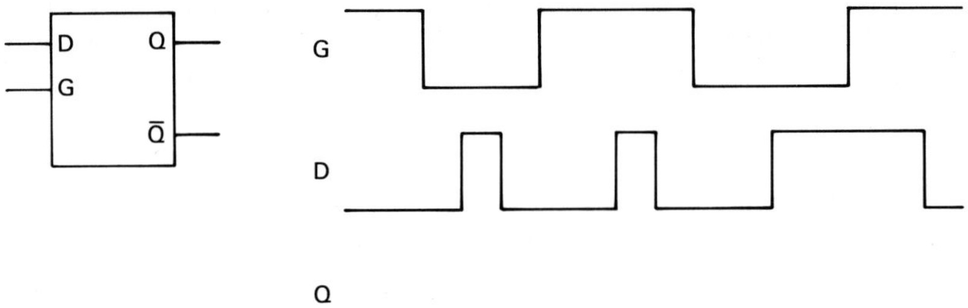

Figure 10-3

11. How many flip-flops are in the 7475 IC?

12. What is the difference between the enable input of the 7475 and the clock input of the 7474?

13. Sketch the Q output waveform for the circuit shown in Figure 10-4.

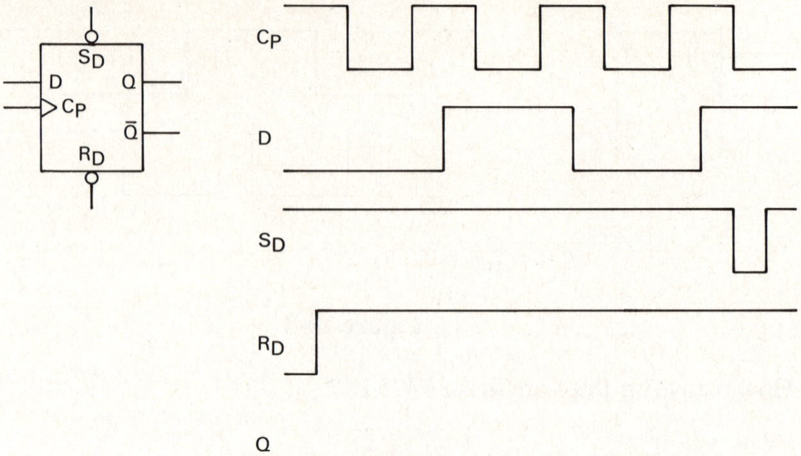

Figure 10-4

14. On a J-K flip-flop, when is the flip-flop in a hold condition?

15. How is a J-K flip-flop made to toggle?

16. On a master-slave flip-flop, when is the master enabled?

17. When will the Q output of the master-slave flip-flop change?

18. What does the triangle on the clock input of a J-K flip-flop mean?

19. Sketch the output waveform for the circuit shown in Figure 10-5.

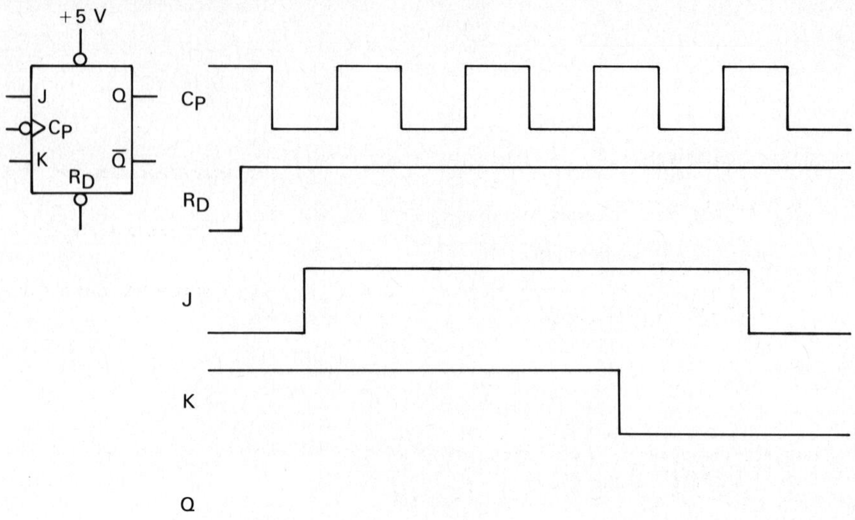

Figure 10-5

20. What is the difference between the 7476 and the 74LS76?

11

CHAPTER

Practical Considerations for Digital Design

Flip-Flop Time Parameters

- A race condition occurs when the data inputs (J-K or D) to a flip-flop are changing at the same time as the active clock transition.
- Set-up time is the time that the data inputs to a flip-flop must be set before the active clock transition.
- Hold time is the amount of time that the data inputs must maintain their level after active clock transition.
- The time required for the Q output to change as a result of some input is called the propagation delay.
  t_{PLH} is the propagation delay when the Q output goes from low to high.
  t_{PHL} is the propagation delay when the Q output goes from high to low.
- The propagation delay can be caused by an asynchronous input (R_D or S_D) or a synchronous input (Cp).
- There are several other parameters that will also be listed in a manufacturers data sheet.
  1. Maximum frcqucncy (F_{max}) - The maximum frequency allowed at the clock input.
  2. Clock pulse width (LOW) [t_W(L)] - The minimum width that is allowed at the clock input during the LOW level.
  3. Clock pulse width (HIGH) [t_W(H)] - The minimum width that is allowed at the clock input during the high level.
  4. Set or Reset pulse width (LOW) [t_W(L)] - The minimum width of the LOW pulse at the set or reset inputs.

- There are times when a delay may need to be introduced in a signal in order for a circuit to function properly.
- Special delay gate ICs are available that provide exact, predefined delays.

Automatic Reset

- A series RC circuit can be used to set or reset a flip-flop when power is applied.
- The voltage across the capacitor will start low and charge toward the supply voltage. This means that the set or reset function will be initially active but will be removed after a short period of time.
- The amount of reset time can be increased by increasing the size of the capacitor.

Schmitt Trigger ICS

- A Schmitt trigger is a circuit used to transform slowly changing waveforms into sharply changing, jitter free output signals.
- The hysteresis is the difference in switching levels between the positive-going input and negative-going input signal.
- The difference in switching voltage is called the hysteresis voltage.
- The transfer function is used to illustrate the relationship between the input switching voltages and the output HIGH and LOW level voltages.

Switch Debouncing

- Any mechanical switch is going to exhibit bounce which will result in multiple pulses.
- One way to debounce a single-pole, single-throw switch is to use a Schmitt trigger circuit with a resistor and capacitor on the input.
- A single-pole, double-throw switch can be debounced using a cross-coupled NAND flip-flop.
- Single-pole, double-throw switches can also be debounced with any type of flip-flop that has an asycnhronous set and reset input.

Sizing Pull-up Resistors

- Pull-up resistors must be sized to maintain the proper voltage levels given the input current requirements of the circuit.
- Pull-down resistors usually are required to dissipate more power.

Practical Input and Output Considerations

- TTL circuits require a supply voltage of 5.0v ±5%.

- A 60 Hz. clock signal can be derived from the AC line voltage because that line frequency is very accurate.
- It is best to connect LEDs to digital outputs so that a low will turn on the LED. TTL can sink more current in the low state than it can drive in the high state.
- Phototransistors are transistors that are used to detect the presence or absence of light.
- Phototransistors combined with waveshaping circuits and flip-flops can be used to make an alarm circuit.
- An optocoupler is an IC with an LED and a phototransistor in the same package.
- Optocouplers can be used to shift or change the output voltage level of a logic circuit.
- Power MOSFETs can be connected to the outputs of logic gates to increase the current drive of the output.

Additional Notes

Unused inputs of logic gates should never be left floating but rather should be connected such that they do not affect the circuit operation. The inputs on AND and NAND should be tied high while the inputs on OR and NOR should be tied low. If the unused input of a logic gate picks up noise the output of the gate will respond to the noise. But the moment the noise goes away the output will return to its normal state. The results of noise pick-up on unused inputs of a flip-flop can be quite a different story. The set, reset and clock inputs can be quite susceptible to noise and should never be left floating. A noise spike on the set or reset input, for example, could change the state of the flip-flop. Once the noise goes away, the flip-flop will not return to its previous state. It is very important that any unused input be connected so that it will not affect the circuit. Low active set and reset inputs, for example, should be tied high.

TEST 11

1. Define a race condition for a flip-flop.

2. What is the difference between set-up time and hold time?

3. Look up the propagation delay from the clock to the output for the 7476. Are the high-to-low and low-to-high propagation delays the same?

4. Is the propagation delay from the clock to the output for the 7476 the same as the delay from the set or reset to the output?

5. Why does the data sheet for the 7476 only give a minimum value for the clock pulse width (both HIGH and LOW)?

6. Why would a delay gate be needed for a digital circuit?

7. What is the range of delay that would typically be needed in a race condition?

8. How much set-up time (ts) is required for the 74LS76?

9. In the automatic reset circuit for a flip-flop, how long does it take the capacitor to completely charge?

10. Can the automatic RC circuit be used to set a flip-flop rather than reset the flip-flop?

11. Why does the hysteresis of a Schmitt trigger prevent input jitter from causing the output to switch back and forth?

12. A Schmitt trigger has V_{t+} = 2.0V and V_{t-} = 1.2v. What is the hysteresis voltage of the Schmitt trigger?

13. Sketch the transfer function for the Schmitt trigger in problem 12. Assume standard TTL voltage level outputs.

14. What is meant by the term switch bounce?

15. Draw a debounce circuit for a single push-button switch using a Schmitt trigger.

16. Show how a 74LS76 can be used to debounce a single-pole, double-throw switch.

17. What would be the output voltage of a 7814 voltage regulator?

18. Why is the Schmitt trigger needed in the 60 Hz TTL-level clock pulse generator?

19. Why should an LED be pulled low from a logic gate rather than pulled high?

20. The output of a standard TTL NAND gate is used to pull an LED indicator low. The LED is in series with a 470 ohm resistor. What is the current in the circuit when the LED is on?

9. A 22 MHz clock signal is put into a MOD-16 counter. What is the frequency of the Q output of each stage of the counter?

10. List which pins need to be connected together on a 7493 to make a MOD-12 counter.

11. What is the difference between a 7490 and a 7492?

12. What is the difference between a 7490 and a 7493?

13. List which pins need to be connected together on a 7492 to make a MOD-12 counter.

14. Which segments (by letter) of a seven segment display need to be active in order to display a digit 6?

15. List the state of each output pin of a 7447 if RBI = 0, LT = 1, A0 = 1, A1 = 0, A2 = 0, and A3 = 1.

16. Why can a synchronous counter operate at a higher frequency than a ripple counter?

17. Sketch a 3-bit synchronous counter using J-K flip-flops.

18. If synchronous counters are better than ripple counters, why are synchronous counters not used all the time?

19. Show the pin connections to make the 74192 a BCD down counter.

20. What is meant by parallel load of a counter?

13

CHAPTER

Shift Registers

Shift Register Basics

- A register acts as a temporary storage device for a group of data bits.
- A shift register is used to move data to the left or to the right by one bit for each input clock pulse.
- Parallel load means to load all flip-flops of a register at one time.
- Serial load means to load the flip-flop of a register one bit at a time.
- The number of clock pulses required to shift all bits of a register completely in or completely out of the register is equal to the number of flip-flops in the register.
- There are four basic register configurations:
  1. Parallel-in, Parallel-out.
  2. Serial-in, Serial-out (both left shift and right shift).
  3. Parallel-in, Serial-out (right shift).
  4. Serial-in, Parallel-out (right shift).
- Many computers operate on parallel data; but, this data must be converted to serial format to be sent over phone lines.
- ICs called UARTs are used to interface microprocessors and parallel data to communications links that use serial data.
- This parallel-to-serial conversion and serial-to-parallel conversion can be performed by shift registers.

Parallel-to-Serial Conversion

- In order to construct a right shift register, the outputs of each flip-flop (Q and $\overline{Q}$) must be connected to the inputs (J and K or D) of the flip-flop to its right.
- A shift register must also have all clock inputs tied to a common clock signal so they will all change state at the same time.
- The asynchronous inputs (set and reset) for each flip-flop can be used to provide the parallel load.

- The output of a right shift register will appear at the Q output of the LSB. The LSB will appear first at the output.
- The output of a left shift register will appear at the Q output of the MSB. The MSB will appear first at the output.

Recirculating Register

- By taking the serial output (Q and $\overline{Q}$ of the LSB) and connecting it back to the serial input (J and K of the MSB) a recirculating register can be constructed.
- In the recirculating register, the original data will not be lost.

Serial-to-Parallel Conversion

- Parallel data can be taken from a register by reading the Q output of each flip-flop.
- Serial data can be applied to a shift register by applying one bit at a time to the MSB input for a shift right register.
- For a shift left register, data must be entered one bit at a time at the LSB.
- A strobe line is used to enable the clock so that the register is shifted only once for each data bit.

Ring Shift Counter and Johnson Shift Counter

- Ring shift counters and Johnson shift counters are used to generate sequential control waveforms.
- The Ring shift counter is a recirculating register in which the serial output is connected back to the serial input.
- The Q outputs of the Ring shift counter will go high for one clock pulse one at a time in sequence.
- The Ring shift counter must be initially set-up with a single flip-flop set and all others reset.
- The initial condition of the Ring shift counter will be shifted through the register on each clock pulse.
- The Johnson shift counter has the recirculating lines crossed such that the inverse of the output data is fed back to the input.
- Each Q output of the Johnson shift counter is a square wave (50% duty cycle).
- Each Q output of the Johnson shift counter is offset or shifted by one clock pulse.
- The Johnson shift counter must be initially reset (all flip-flops zero).

Shift Register ICs

- The 74164 is an 8-bit serial-in, parallel-out shift register.
- The 74164 has two data input, one of which can be used as a high active enable.

- The 74164 also has a master reset for all flip-flops.
- The 74165 is an 8-bit serial or parallel-in, serial-out shift register.
- The 74165 has a low active parallel-load terminal ($\overline{PL}$).
- The 74165 also has a chip enable ($\overline{CE}$) for hold of data or enabling the shift. The chip-enable can also be used as a strobe input for serial loading of data.
- The 74194 is a 4-bit bidirectional universal shift register.
- The 74194 is capable of shift-left, shift-right, parallel-in, parallel-out, serial-in, or serial-out.
- The 74194 has two mode control inputs (S0 and S1) which are used to select the desired operating mode.
- A three-state output can be HIGH, LOW or a float.
- When the three-state circuit is enabled, the output is the same as the input. When the circuit is disabled, the output is "high-impedance" or floating.
- The three-state buffers are used when connecting the output of more than one register together to a single input.

System Design Applications for Shift Registers

- A Ring shift counter can be used to generate accurately timed sequence pulses.
- A Johnson ring counter can be used to generate pulses of different lengths and occurring at different times.
- Shift registers can be used to convert serial data to parallel data or to convert parallel data to a serial bit string.

Driving a Stepper Motor with a Shift Register

- Stepper motors make their rotation in discreet steps. This stepping action is caused by digital levels on the motor inputs.
- Shift registers make excellent stepper motor drivers.
- The 74194 is a 4-bit register that can be parallel loaded with the correct binary pattern and then shifted left or right to determine the direction of rotation.
- The clock frequency of the shift register will determine the speed of the motor's rotation.

Three-State Buffers, Latches and Transceivers

- A group of flip-flops is called a register and function as a buffer, a latch or a transceiver.
- The three-state buffer can be used to interface several device outputs to one set of inputs.
- A buffer will pass a digital bit from its input to its output unchanged when the buffer is enabled.

- The 74LS244 is an octal (8) three-state buffer.
- A latch is a flip-flop that can be used to remember digital data.
- An 8-bit latch can be used to interface the output of a microprocessor to other devices.
- The 74LS373 octal latch and the 74LS374 octal D flip-flop are popular microprocessor interface chips.
- A transceiver is a bidirectional buffer.
- With a transceiver, a microprocessor can receive data from an I/O device or can send data out to an I/O device over the same set of common lines (data bus).
- The 74LS245 octal 3-state transceiver uses two 3-state buffers for each set of data lines.

Additional Notes

An important application of the shift register is a delay circuit for a bit string. The time that the bit takes to propagate through the shift register is a function of both the clock frequency and the number of flip-flops in the register. If a clock signal of 10 usec is used to clock an 8-bit shift register than it will take a total of 80 usec (8×10 usec) for a bit at the input of the register to appear on the output. In addition, the shift register can be tapped at each Q output as a function of multiples of the clock frequency. The first Q output will be delayed by 10 usec, the second by 20 usec, the third by 30 usec, and etc. A series of data bits applied to the input of a shift register can then be delayed in the shift register by multiples of the clock frequency.

TEST 13

1. What is meant by parallel load of a shift register?

2. How many clock pulses will be required to completely load serially a 5-bit shift register?

3. Sketch the Q output waveforms from a 4-bit shift right register for four clock pulses if the register is loaded with a hexadecimal 6.

4. What is the difference between a shift right register and a shift left register?

5. How can data be taken parallel out of a shift register?

6. What is a recirculating register?

7. How is a strobe signal used when serially loading a shift register?

8. What is the preset condition for a Ring shift counter?

9. Sketch the output waveforms for a 4-bit Ring shift counter.

10. What is the difference between a Ring shift counter and a Johnson shift counter?

11. Sketch the output waveforms for a 4-bit Johnson shift counter.

12. Show how to connect the 74165 IC as an 8-bit Ring shift counter.

13. Show the necessary connections to preload the 74194 with a hexadecimal D.

14. What are the three output conditions of a three-state buffer?

15. Where is it important to use a three-state buffer?

16. What does the output enable do on the 74395A chip?

17. What is the function of a buffer circuit?

18. Explain what a microprocessor data bus is.

19. Where would a latch circuit be used in a microprocessor system?

20. What is a transceiver circuit?

14

CHAPTER

Multivibrators and the 555 Timer

Multivibrators

- A multivibrator is a circuit that generates a digital waveform either continuous or on demand from some external trigger source.
- A bistable multivibrator has two stable states. A S-R flip-flop is a bistable multivibrator.
- The astable multivibrator is a free-running oscillator. It has no stable states.
- The monostable multivibrator, also known as a one-shot, provides a single output pulse of a specific time length when it is triggered from an external source.
- In the astable and the monostable multivibrator RC circuits are used to provide specific timing pulses.

Capacitor Charge and Discharge Rates

- A capacitor will charge exponentially toward the value of supply voltage. The rate that the capacitor charges is dependant on the product of R times C which is known as the time constant of the circuit.
- In most multivibrator circuits the capacitor starts with an initial charge.
- The voltage across the capacitor is exponential no matter if it is charging or discharging.
- The basic charging capacitor formula is:

$$\Delta v = E(1 - \varepsilon^{-t/RC})$$

- Think of Δv as the distance that the capacitor voltage travels and E as the total distance that the capacitor voltage is trying to travel.
- Solving the basic formula for time gives the formula:

$$t = RC \ln(1 + 1 - \Delta v/E)$$

- Both of these formulas can be used for a capacitor either charging or discharging.

Astable Multivibrator

- One astable multivibrator can be built using a Schmitt trigger and an RC circuit.
- The capacitor will charge and discharge between the positive threshold (V+) and the negative threshold (V-) of the Schmitt trigger.
- The output of the Schmitt trigger will be a pulse waveform between V_{OH} and V_{OL} of the chip being used.

Monostable Multivibrator

- The monostable multivibrator (one-shot) is normally in the reset state and will change to the set state when triggered by the input pulse.
- The input trigger can be either positive-edge triggered or negative-edge triggered depending on the one-shot circuit.
- The output pulse width from the one-shot (t_w) is determined by an RC circuit.
- A very simple one-shot can be built from two NAND gates and a RC circuit.
- The output pulse width depends upon the RC circuit and the input and output voltage requirements of the chip being used.

IC Monostable Multivibrators

- The 74123 and the 74121 are both monostable multivibrators. The R and C timing components must be connected externally.
- The 74121 has two active-LOW and one active-HIGH trigger inputs.
- The 74121 has a built-in 2 k-ohm timing resistor or an external resistor can be used.
- The pulse width of the output can be calculated from the formula $t_w = R_{ext}C_{ext} \ln 2$.
- The ICs can be cascaded to provide specific pulse widths at a delay time from the trigger input pulse.

Retriggerable Monostable Multivibrator

- The 74121 is non-retriggerable which means that if a second trigger pulse occurs before the end of the pulse width, it will be ignored by the circuit.

- The 74123 is a retriggerable one-shot. If a second trigger pulse occurs before the end of the timing pulse the 74123 will start the timing cycle all over again.
- The 74123 is also a dual one-shot (two in the same Package) and it has an active-LOW reset to terminate the timing function.
- The pulse width of the 74123 is determined by the following formula:

$$t_w = 0.28\ R_{ext}\ C_{ext}\ (1 + 0.7/R_{ext}).$$

Astable Operation of the 555 IC Timer

- The 555 can be connected as an astable oscillator or as a one-shot.
- A comparitor is a circuit that will compare two analog voltages. The output will be HIGH if the + input is more positive than the - input. The output will be LOW if the + input is less positive than the - input.
- The 555 can be connected as an astable oscillator (see Figure 14-20 in the textbook).
- The external capacitor will charge through $R_A + R_B$ to 2/3 V_{cc}.
- The external capacitor will discharge through R_B to 1/3 V_{cc}.
- The discharge time (output LOW) can be calculated by the formula

$$t_{LO} = 0.693R_BC.$$

- The charge time (output HIGH) can be calculated by the formula

$$t_{HI} = 0.693(R_A + R_B)C.$$

- A 50% duty cycle can be achieved if R_B is bypassed with a diode and $R_A = R_B$.

Monostable Operation of the 555 IC Timer

- The 555 can be connected as a one-shot multivibrator (See Figure 14-24 in the textbook).
- A negative going trigger input is applied to pin 2 and the output is taken from pin 3.
- The pulse width of the output pulse is determined from the formula

$$t_w = 1.10R_AC.$$

- The output of the 555 one-shot is a positive-going pulse.

Crystal Oscillator

- A quartz crystal oscillator is extremely more accurate than using an RC timing circuit.
- A quartz crystal oscillator is very stable compared to an RC timing circuit.
- In a voltage-controlled oscillator (VCO) the output frequency will change as the input voltage changes.

Additional Notes

A voltage-controlled oscillator will change output frequency if the input voltage is changed. There are many ICs which are designed as VCOs but the 555 timer circuit can also be used as a VCO. The Control Voltage input of the 555 connects to the 2/3 V_{cc} point which is the reference point for the reset comparitor. By applying an external voltage to this pin then the reference voltage for the reset comparitor will be changed. Changing this voltage changes the amount of time it takes the capacitor to discharge and reset the flip-flop. If the discharge time is changed then the period of the output waveform is changed and thus the frequency will change. To make the 555 a VCO, it must first be configured as an astable multivibrator and a control voltage can be applied to the Control Voltage input (pin 5).

TEST 14

1. What is another name for a bistable multivibrator?

2. What is the difference between an astable multivibrator and a monostable multivibrator?

3. What controls the output pulse width of a one-shot?

4. A 22 k-ohm resistor and a .02 uF capacitor are connected in series to a 5 V source. How long will it take the capacitor to charge to 3.4 volts?

5. A 0.3 uF capacitor has an initial charge of 2.7 volts. If it is discharged through a 10 k-ohm resistor for 6.5 msec, how much voltage will be left across the capacitor?

6. If the resistor in the Schmitt trigger astable multivibrator is a variable resistor, what about the output voltage waveform will change when the resistance is changed?

7. A 25 uF capacitor is connected between pins 9 and 10 of a 74121 one-shot. What would the output pulse width be?

8. Design a circuit using 74121 to convert a 33 kHz, 30% duty cycle waveform to a 33 kHz, 60% duty cycle waveform.

9. What is the difference between a retriggerable one-shot and a nonretriggerable one-shot?

10. Is there any limit to the number of times that a 74123 can be retriggered?

11. Does the 74123 one-shot have an internal timing resistor?

12. What is the pulse width of the output of a 74123 if an external 4.7 k-ohm resistor and .022 uF capacitor are used?

13. What is the function of the comparitors in the 555 timer circuit?

14. What does the discharge transistor do in the 555 timer circuit?

15. If V_{cc} of a 555 timer circuit is set to + 10 V what is the level of output voltage from the circuit?

16. Determine t_{HI} and t_{LO} for the circuit of Figure 14-1.

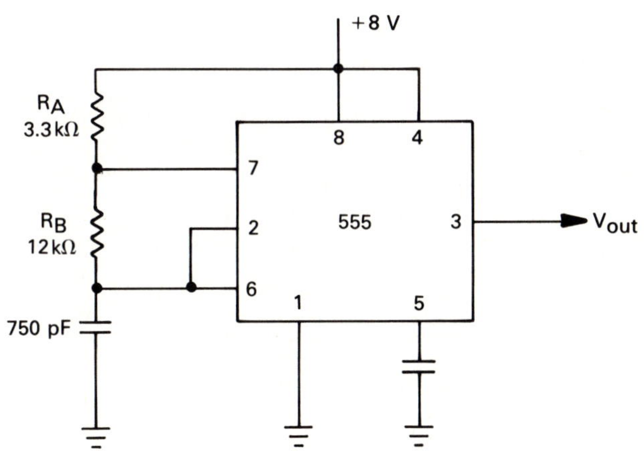

Figure 14-1

17. What is the duty cycle of the waveform at the output of the circuit in Figure 14-1?

18. If a diode is connected across resistor R_B (Positive end up) in Figure 14-1 what is the new duty cycle of the output waveform?

19. What is the output pulse width of the waveform at the output of the circuit in Figure 14-2?

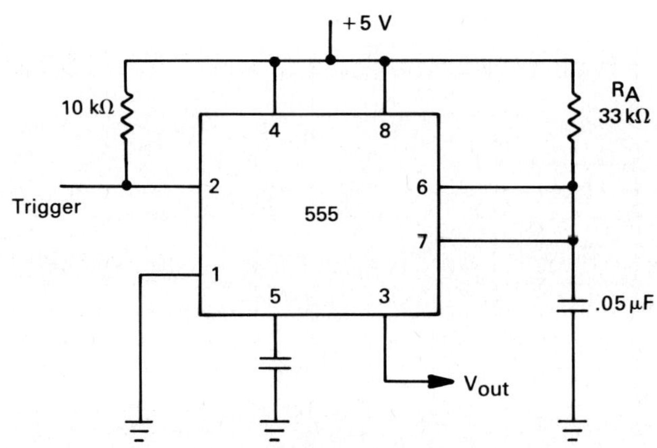

Figure 14-2

20. Is the circuit of Figure 14-2 an astable multivibrator or a monostable multivibrator?

15

CHAPTER

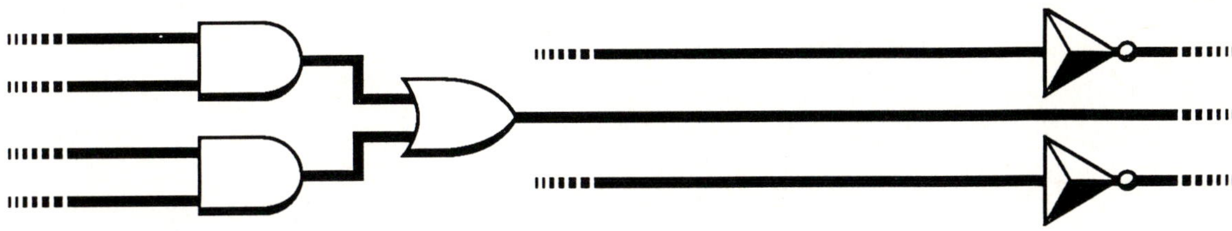

Interfacing to the Analog World

Digital and Analog Representations

- Analog quantities can be represented by digital numbers.
- A range of the analog quantity can be divided into small steps where each step is represented by a digital number.
- The resolution is the number of bits in the digital number and will determine the number of steps in the analog range.
- The number of steps is 2^N where N is the number of digital bits.

Operational Amplifier Basics

- An op amp is a special amplifier circuit that has very high input impedance, very high gain and very low output impedance.
- Since the input impedance is very large, no current can flow into the op amp.
- There is no voltage difference between the inverting input (-) and the non-inverting input (+). This keeps the inverting input (-) at virtual ground.
- The current in the input resistor (R1) is equal to the current in the feedback resistor (R_F) and the output voltage is the feedback current (I_F) times the feedback resistance.
- If there is more than one input resistance then the feedback current is equal to the sum of the input currents.

Binary-Weighted Digital-to-Analog Converters

- Each input resistance of a binary-weighted DAC must be weighted to the binary input it represents.

- Starting with the LSB each resistor at the next higher input is 1/2 the resistance of the previous value.
- The current in the feedback resistor will be the sum of the input currents. Therefore, the output voltage will be proportional to the binary number represented on the inputs.
- The problem with this type of DAC is that it takes a large number of resistor values for large binary inputs.

R-2R Ladder Digital-to-Analog Converters

- The R-2R ladder requires only 2 different values of resistors no matter how many digital bits are on the input.
- The voltage contributed by each binary input is 1/2 the previous input. The MSB contributes a voltage equal to the input reference. Each input thereafter is 1/2 the value of the input before it.
- If the binary input to any DAC goes through its natural binary count sequence then the output of the DAC is a staircase waveform. Each step represents another count of the binary input.

Integrated-Circuit Digital-to-Analog Converters

- The DAC0808 is a very popular 8-bit DAC that uses an R-2R ladder converter.
- The reference current on the reference input will determine the maximum current on the output.
- The output current should be used to drive an op amp in order to maintain the accuracy of the conversion.
- A DAC can be tested by connecting a counter to the input and observing the staircase waveform at the output on an oscilloscope.

IC Data Converter Specifications

- There are many specifications that are important in the selection of DAC and ADC.
- Some of the more important specifications include:
  differential non-linearity, also called step size error.
  gain error which is the wrong slope to the staircase waveform.
  a missing code which may be the result of a fault in the circuit.
  offset error occurs when the entire staircase waveform is shifted.

Parallel-Encoded Analog-to-Digital Converters

- The parallel-encoded ADC is also known as simultaneous, multiple comparitor and flash converter.
- Several comparitor circuits are used and each one is referenced to a different voltage through a voltage divider.
- The output of the comparitors are fed to a priority encoder in order to get the binary number as an output.
- The outputs of the comparitors will depend upon the input voltage. The larger the input voltage the more comparitors that will have high outputs.
- The priority encoder will code into binary the value of the highest active input.
- The parallel-encoded ADC operates very fast but requires a large number of comparitors. $2^N - 1$ comparitors are required where N is the number of binary bits.

Counter-Ramp Analog-to-Digital Converters

- The counter-ramp method uses a binary counter, a D/A converter, and a voltage comparitor.
- The counter starts at zero. The output of the D/A converter is a staircase waveform that increases in voltage with each increment of the counter. This continues until the staircase waveform equals or exceeds the input analog voltage. This will stop the counter and display the count on the output.
- The counter ramp is simple to build but it operates very slowly. Maximum conversion time is (clock period) (2^N) where N is the number of bits.

Successive-Approximation Analog-to-Digital Conversion

- The successive-approximation method is the most popular method of ADC. A successive-approximation register is used in place of a counter.
- The successive-approximation register will zero in on the correct binary number by testing each binary bit one at a time starting with the MSB.
- The successive-approximation method is much faster than the counter-ramp method. N clock pulses are required for conversion rather than 2^N clock pulses.

Integrated-Circuit Analog-to-Digital Converters

- The NE5034 is an 8-bit successive-approximation converter with a three-state buffered output.
- The ADC0801 is also an 8-bit successive-approximation converter. It is compatible with microprocessor systems with control signals such as chip select ($\overline{CS}$), Read ($\overline{RD}$), Write ($\overline{WR}$) and output signals like interrupt ($\overline{INTR}$).

- The ADC0801 also has separate grounds for the analog and the digital systems.

Data Acquisition Systems

- DAS are used to accumulate a number of analog signals for microprocessor systems.
- The communication between the microprocessor and the components of the DAS is called handshaking or polling.
- The analog multiplexer selects one of the eight input signals to be sampled.
- The sample and hold circuit takes a quick sample of the analog voltage and holds that value until the next sample time.
- The programmable-gain amplifier can provide a different gain for each of the different input signals so that they will all be to the same scale.
- The ADC converts the final analog voltage to a digital number that can be processed or stored by the microprocessor.

Transducers and Signal Conditioning

- A transducer converts a physical quantity such as heat, light or force to an electrical quantity.
- Signal conditioning is necessary to change the electrical output of a transducer to a uniform, linear signal of a constant range.
- A thermistor changes resistance with changes in temperature. The thermistor is a non-linear device.
- The thermistor can be used as part of an op amp circuit that will provide an output voltage that changes with temperature.
- There are other devices, such as the LM35 and the LM185, which will produce a linear change with respect to temperature.
- A strain gauge will change resistance when the device is stretched or elongated. This means that the device can measure physical strain.
- The change in resistance of a strain gauge is linear but very small. Therefore, sensitive instrumentation amplifiers must be used to convert the change in resistance to a change in voltage.

Additional Notes

A dynamic test of a digital-to-analog converter can be made by connecting a counter to the input of the DAC and observing the staircase waveform at the output with an oscilloscope. The staircase waveform can indicate a number of problems with the DAC circuit. A step size error or gain error (change in the slope) may indicate a changed value of resistor or a drift in the amplifier parameters. An offset error could be a result of a change in input conditions. A loss of data bits on the input or a failure of part of the input circuit would cause missing code error or may cause the output to be non-monotonic. In order for the output to be monotonic, the staircase should increase with uniform steps and no downward steps. Faulty inputs or bad components within the DAC can cause the output to step out of sequence. Many of these errors can be detected and identified by careful analysis of the output staircase waveform.

TEST 15

1. How many different voltages can be output from a DAC with a 6-bit resolution?

2. If the range of output voltage of a 6-bit DAC is 0 to 15 volts, what is the step voltage of the output?

3. What is the current in R1 and the current in R2 for the circuit shown in Figure 15-1?

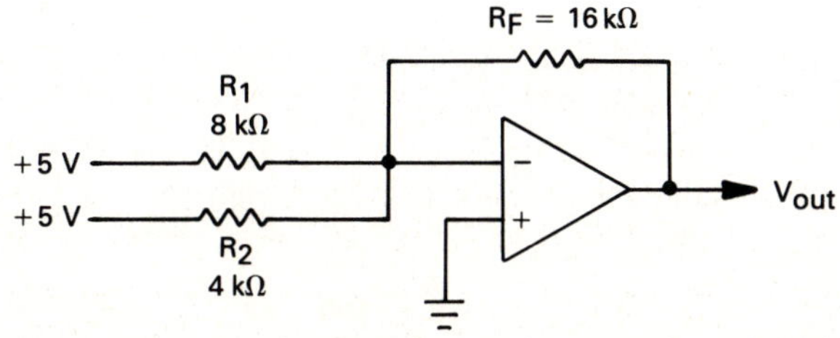

Figure 15-1

4. What is the current in the feedback resistor for the circuit of Figure 15-1?

5. What is the output voltage for the circuit shown in Figure 15-1?

6. Why is a binary-weighted DAC usually limited to 4-bit binary conversion?

7. What type of DAC is shown in Figure 15-2?

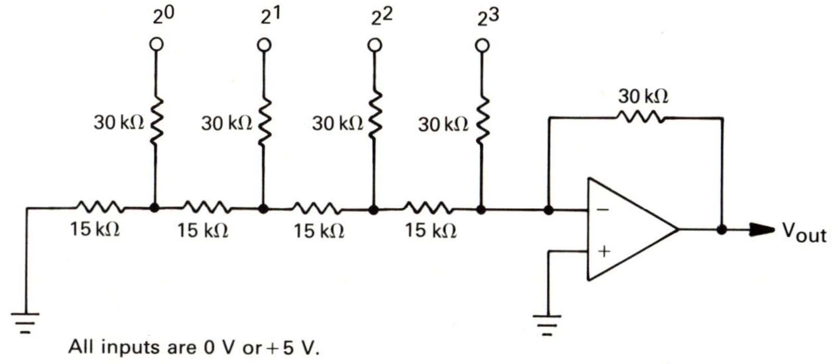

Figure 15-2

8. What is the output voltage of the circuit of Figure 15-2 if the inputs are as follows: $2^0 = 1$, $2^1 = 0$, $2^2 = 1$, $2^3 = 0$?

9. What is the maximum output voltage for the circuit shown in Figure 15-2?

10. What is gain error in a DAC?

11. Describe offset error for a DAC.

12. What is one advantage to using a parallel-encoded (flash) ADC?

13. If an analog signal is to be converted to an 8-bit resolution, how many comparitors are used in a parallel-encoded ADC?

14. If the same analog signal is to be converted to an 8-bit resolution using a counter-ramp ADC, how many comparitor circuits would be used?

15. What is the disadvantage to using a counter-ramp type ADC?

16. What does a successive-approximation register do?

17. What is the maximum conversion time for a counter-ramp ADC with 8-bit resolution and a clock frequency of 20 kHz?

18. What is the maximum conversion time for an 8-bit successive-approximation ADC with a clock frequency of 20 kHz?

19. Why does the NE5034 have a three-state buffered output?

20. Where would a data acquisition system be used?

16

CHAPTER

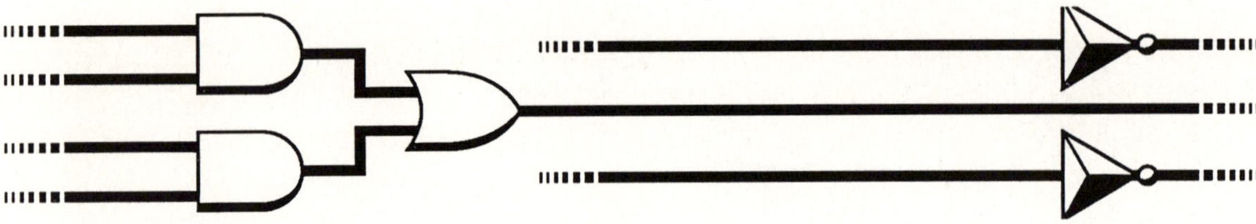

Semiconductor Memory and Programmable Arrays

Memory Concepts

- A memory is a semiconductor or magnetic device used for storage of digital data.
- A memory location is a group of storage devices that will hold one data word.
- A data word length of 8-bits is called a byte.
- Each memory location can store a different data word and has a unique address.
- The address bits must be decoded to select the proper memory location.
- When new data are to be stored in a memory location a "write" operation will be performed.
- To receive the data that have already been stored in a memory location, a "read" operation will be performed.
- Timing diagrams are used to illustrate the proper operation of the memory cycle.
- Address lines and data lines are usually grouped together on a timing diagram.
- The timing diagram will show the proper sequence of pulses, proper polarities, setup time, pulse widths, and any propagation delays.
- During a read operation the three-state output must be enabled for the selected memory location only.
- If more than one output is enabled at a time, there will be a bus contention.

Static RAMs

- Random Access Memory (RAM), also known as read/write memory, can access data at any selected memory location.
- The static RAM uses flip-flops as the basic storage element or cell and can retain the stored data as long as power remains applied to the memory.
- The dynamic RAM uses capacitors as the basic storage element and must be refreshed on a regular basis in order to retain the data in the memory.
- Dynamic RAMs are more dense than static RAMs and their cost per bit is less than the static RAMs.
- The 2147H is a 4K × 1 MOS static RAM.
- The 12 address lines of the 2147H are separated into two groups of six. One group of six address lines is decoded into 64 row select lines and the other group is decoded into 64 column select lines.
- The input data for the 2147H are ANDed with the $\overline{CS}$ (LOW) and the $\overline{WE}$ (LOW). The output data bit are ANDed with the $\overline{CS}$ (LOW) and the $\overline{WE}$ (HIGH).
- For a read operation, a valid address is presented at the same time as $\overline{CS}$ goes low. The $\overline{WE}$ must be high. A short period of time later, called the access time, the output data are valid.
- During a write operation the input data must be valid before $\overline{CS}$ or $\overline{WE}$ returns HIGH and must remain valid for a time after $\overline{CS}$ or $\overline{WE}$ return HIGH.
- Each 2147H chip can store 1 bit of data at a time. To build a memory with an 8-bit word, 8-2147H chips would be required.
- The address lines and the control lines ($\overline{CS}$ and $\overline{WE}$) would be connected in parallel to all 8 chips.
- This will leave 8 separate data input lines and 8 separate data output lines for the 4K × 8 memory system.
- Many other memory chips with different configurations are also available.

Dynamic RAMs

- Dynamic RAMs use capacitors for each cell rather than flip-flops. These dynamic RAMs must be refreshed every 2 ms or less to maintain the charge on the capacitor.
- The Intel 2118 chip is an example of a 16K × 1 dynamic RAM.
- The 7 address lines of the 2118 must be multiplexed from a total of 14 address lines. The $\overline{RAS}$ (Row Address Select) and the $\overline{CAS}$ (Column Address Select) leads are used to help access the complete 16,384 addresses.
- Other than the addition of the $\overline{RAS}$ and the $\overline{CAS}$ leads, the read cycle and write cycle timing is very similar to the static RAM.
- There are three ways that the dynamic RAM can be refreshed:
  1. perform a read cycle
  2. perform a write cycle
  3. perform a $\overline{RAS}$-only cycle
- All cells or 128 rows must be refreshed every 2 ms or less to ensure data retention.

- There are special controller ICs to handle the demultiplexing and refreshing of dynamic RAMs. An example of such an IC is the Intel 3242.
- The 3242 will multiplex 14 address lines to 7 lines for the dynamic RAM.
- The 3242 is capable of doing a "burst" refresh of all 128 rows. The 3242 has an internal counter that, when enabled, will sequence through all the rows and will output a $\overline{\text{ZERO}}$ signal when the refresh is complete.
- The timing signals for $\overline{\text{RAS}}$, $\overline{\text{CAS}}$, and row enable can be developed using a digital delay line.

Read Only Memory

- The read only memory (ROM) is used to store data on a permanent basis. They are nonvolatile. That is, the data will not be lost when the power is removed from the memory.
- Mask ROMs are available in large quantities. The manufacturer must custom build the chip and the initial start-up cost is very expensive. In large quantities, however, these memories can be quite inexpensive.
- The fusible-link PROM is user programmable. Initially all cell outputs are a logic 0. The user can program a cell for a logic 1 by forcing excessive current to blow the fusible link. Once a cell has been programmed for a 1, it cannot be changed back.
- The erasable PROM (EPROM) can be erased by exposing the chip to ultra-violet light for a period of time. Once the EPROM has been erased it can be reprogrammed with new data.
- The electrically erasable PROM (EEPROM) can be erased while the chip is in the circuit by applying a high voltage to the chip.
- The 2716 is an EPROM organized as a 2K × 8 memory. To read the chip, the correct address is applied and $\overline{\text{CS}}$ and $\overline{\text{OE}}$ are brought low. The data will appear on the output pins.
- After the 2716 has been erased by a strong UV light, it can be reprogrammed by setting V_{pp} = 25 v (program voltage) and $\overline{\text{OE}}$ HIGH. Set the proper address on the address pins and the data to be stored on the output data pins. Finally, pulse $\overline{\text{CE}}$/PGM HIGH for 50 ms to store the data.

Memory Expansion and Address Decoding

- When the size of a desired memory system is larger than the number of locations in the chip being used, then several chips will have to be used to provide the additional memory.
- These additional memory chips must be selected (enabled) with additional address decoding. Most microprocessors will provide up to 16 address lines (64K).
- The LSBs of the address bus are connected directly to the memory chips while the higher order address leads are connected to additional decoder chips.

  The outputs of these decoders are used to enable individual memory chips.
- Additional decoders can be connected to the address bus to provide memory for all 64K memory addresses.

- Many memory chips (especially MOS devices) require that the data output lines be buffered in order for them to drive heavy loads.

Programmable Logic Devices

- Large sum-of-product Boolean Expressions can be implemented by using programmable logic devices (PLDs).
- PLDs are available as Programmable Read-Only Memory (PROM), Programmable Array Logic (PAL), and as Programmable Logic Arrays (PLA).
- The standard PAL has multi-input AND gates connected to OR gates and inverters. The inputs have fusible-links so that the logic can be custom built by the user.
- A hard array logic (HAL) requires a custom mask made by the manufacturer.
- PALs have fixed OR gates and PLAs have programmable OR gates.
- These PLDs use a one line convention for showing the logic connections.
- Many of the newer arrays are also available with sequential logic (D and S-R flip-flops) available in the IC.
- In order to program a field programmable logic array (FPLA), a special programming system that interprets the data and blows the correct fuses must be used.
- In order to use the programming system a programming table must be filled out.

Additional Notes

There are many applications for Read Only Memory (ROM). The most common of these is the program storage in microprocessor systems. This application is sometimes called firmware. Another application of the ROM is in implementing combinational logic. Once a ROM is programmed, the output pattern will be the same every time a given address is applied. This relationship between the address and the output data can be tabulated in a truth table. This truth table can look just like the truth table for a combinational logic circuit. In fact, given the desired truth table, a ROM can be programmed to provide the proper sequence of output leads for the combinations of inputs. This means that the ROM can replace a large combinational logic circuit.

TEST 16

1. How many address lines would be required for a 2K × 4 memory chip?

2. Describe the timing diagram of a write operation.

3. What is meant by the setup time for a memory write?

4. What is the difference between static RAM and dynamic RAM?

5. What does the term random access mean in terms of a memory?

6. How many 1K × 4 RAM chips would be required to build a 1K × 8 memory system?

7. Why is a refresh cycle necessary for a dynamic RAM?

8. What is the maximum time required before a dynamic RAM must be refreshed?

9. Briefly describe the function of a dynamic RAM controller.

10. Why do most dynamic RAMs use a multiplexed address bus?

11. What is a multitap digital delay line?

12. Which type of ROM has to be custom built by the factory?

13. Which type of ROM can be erased by UV light?

14. Which type of ROM can be erased by an electrical signal?

15. Why are ROMs called nonvolatile memory?

16. Why do most ROMs have an output enable as well as a chip enable?

17. How many 2K × 8 ROM chips would be required to build a 16K × 8 memory system?

18. Could a 74LS138 decoder be used to decode the memory chips in problem 17?

19. What is the difference between a PLA and a PAL?

20. What is the difference between a HAL and a PAL?

17

CHAPTER

Microprocessor Fundamentals

Introduction to System Components and Buses

- The 8085 is a popular 8-bit microprocessor that is controlled by software programs that allow it to do arithmetic, logic, and I/O operations.
- The address bus is a group of 16 common connections that allow the microprocessor to select a particular memory location or I/O device.
- The data bus is an 8-bit bidirectional bus that allows the microprocessor to communicate (exchange data) with memory or with I/O devices.
- The control bus is of varying width and is a set of connections that the microprocessor can use to tell other ICs in the system whether to do a read operation or a write operation or a memory access or an I/O access.
- The address decoder is an IC or group of ICs that decode the address sent by the microprocessor and enable the correct memory chip. The address decoder enables only one memory IC at a time.
- Any microprocessor based system will have two types of memory: RAM and ROM. Permanent information is stored in ROM and temporary information is stored in RAM.
- An input port will provide data to the microprocessor from an input device such as a keyboard.
- An output port is a device where the microprocessor can send data that must later be sent to an output device such as a printer.

Software Control of Microprocessor Systems

- The big advantage to microprocessor based systems is that changes or modifications to the system can be done by simply changing the software. The programs that define the sequence of events for the microprocessor are called software.
- An internal register in the 8085 microprocessor that is used to exchange data with an I/O device is called the accumulator.
- Microprocessors are programmed directly in assembly language.
- Most assembly language commands contain a mnemonic and an operand. The mnemonic is a symbolic spelling of the instruction and the operand is the address or data that the instruction is to work with.
- All microprocessors have similar sets of commands.

Internal Architecture of the 8085A Microprocessor

- The 8085A microprocessor is an improved version of its predecessor the 8080.
- The 8085A has an 8-bit accumulator and six general purpose 8-bit registers: the B, C, D, H, and L.
- All arithmetic and logic operations are done in an ALU and the status of the results is held in a flag register.
- As instructions are read from memory, they are placed in the instruction register and decoded in the decoder.
- The 8085A also has an interrupt structure for external devices to get the microprocessor attention and a serial input and output port for exchange of serial data.
- In addition to the six general purpose registers, the 8085A also has three 16-bit registers: the stack pointer, the program counter, and the address latch.

Instruction Execution Within the 8085A

- Mnemonic codes cannot be understood by the computer. They must first be assembled or converted to binary machine codes.
- Machine language codes are first read from RAM into the instruction register.
- The instruction register passes the code to the instruction decoder which interprets the instruction.
- If address or data follows the command, then the microprocessor will read the next memory location until it has all the data or address.
- The final step is for the microprocessor to perform the function called for by the instruction command.

Hardware Requirements for Basic I/O Programming

- In memory mapped I/O systems each I/O device is seen as a memory location and uses a 16-bit address.
- Standard I/O or I/O mapped systems use a separate 8-bit port address and special instructions to get to the I/O devices.
- Standard I/O requires additional hardware to decode the 8-bit address for device selection.
- The special I/O instructions are the IN and the OUT instruction.
- In addition to the port address decoder, an input port uses tri-state buffers to interface the data bits onto the microprocessor data bus.
- An output port must include an 8-bit latch as well as the 8-bit address decoder.

Writing Assembly Language and Machine Language Programs

- Software instructions written in mnemonic abbreviations must first be converted to machine language codes.
- Converting mnemonic abbreviations to machine codes can be done by hand or it can be done by a computer program called an assembler.
- Higher level languages like Pascal, BASIC, and FORTRAN are much easier to write but are not as efficient or as fast as assembler language.
- Assembly language is different for each microprocessor.

Survey of Microprocessors and Manufacturers

- The first microprocessor was a 4-bit machine.
- The first 8-bit machine was introduced by Intel in 1974.
- Many manufacturers have entered the microprocessor race; however the two major manufacturers have been Intel and Motorola.
- A microcontroller is a chip with a microprocessor, RAM, ROM, and I/O all on one chip.
- The microcontroller is very popular for many control applications.

Additional Notes

An important role in software development has been the software development tools. Systems have been developed that aid the programmer in developing an entire microprocessor based system. These development systems usually contain CRT displays and full ASCII keyboards for program and data entry. They usually have disk drives for storage of programs. They have printers to make hard copy printouts. And, they generally have the hardware necessary to burn programs directly into ROMs and EPROMs.

The development system also will have a number of software packages to aid the programmer as well. These systems usually include assemblers to convert mnemonic code directly into machine code. They usually have high level language compilers. They generally contain the ability to link several programs or sub-programs together into one larger program. And, they usually contain some sort of diagnostic or testing ability so that the programmer can determine if the program is going to run correctly before he ever puts it into the microprocessor. These development systems are usually very expensive; however they make the job of software development much easier.

TEST 17

1. How many buses are connected as part of the 8085A microprocessor?

2. How many bits are used in the data bus?

3. How many bits are used in the address bus?

4. Which bus is a bidirectional bus?

5. The items that you can physically touch in a computer system are called:
   a. software
   b. firmware
   c. hardware
   d. none of these

6. The register in the 8085A that is used to keep track of the memory address of the next op-code to be run in the program is the:
   a. stack pointer
   b. program counter
   c. instruction pointer
   d. accumulator

7. The technique of assigning a memory address to each I/O device in the computer system is called:
   a. memory mapped I/O
   b. ported I/O
   c. dedicated I/O
   d. wired I/O

8. Single bit indicators that may be set or cleared to show the results of logical or arithmetic operations are the:
   a. flags
   b. registers
   c. monitors
   d. decisions

9. All computer programs for a machine are called:

10. The devices that provide the means for a computer to communicate with the user or other computers are referred to as:

11. The circuits in the 8085A that provide the arithmetic and logic functions are called the:

12. A register in the microprocessor which keeps track of the answer or results of any arithmetic or logic operation is the:

13. The two instructions on the 8085A that allow the user to specify dedicated or ported I/O are: and:

14. What type of circuit is used at the interface point of an input port?

15. What type of circuit is used at the interface point of an output port?

16. What is the difference between a mnemonic code and machine code?

17. What kind of computer program is used to convert mnemonic code to machine code?

18. How is a high level language program different from a machine language program?

19. When was the first 8-bit microprocessor introduced?

20. What are the two major manufacturers of microprocessors?

Appendix A

ANSWERS TO TEST 1

1. 11
2. 154
3. 153_8
4. 274_8
5. $5D.5_{16}$
6. $CB.C_{16}$
7. 00100011_2
8. 10111011_2
9. 15
10. 29
11. $7D_{16}$
12. $4A_{16}$
13. 111110_2
14. 001000100_2
15. 181
16. 463
17. 431_8
18. 47_8
19. 11001001_2
20. 000101001011_2
21. 569
22. 803
23. 000100100111
24. 010001101001
25. STUDY HARD

ANSWERS TO TEST 2

1. 0.8 MHz
2. 64.9 usec
3. Timing diagram
4. 5.33 usec
5. 0.444 usec
6. 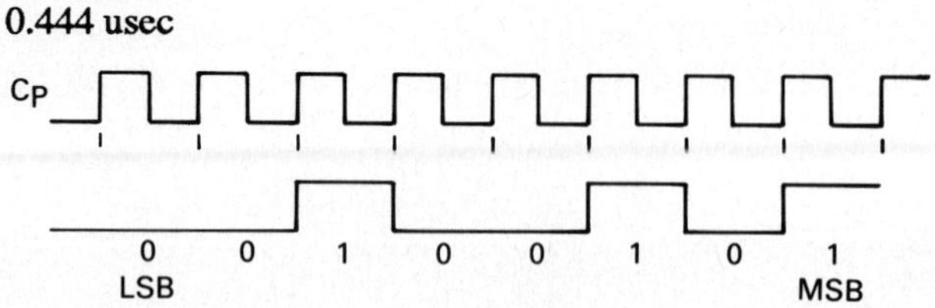

7. 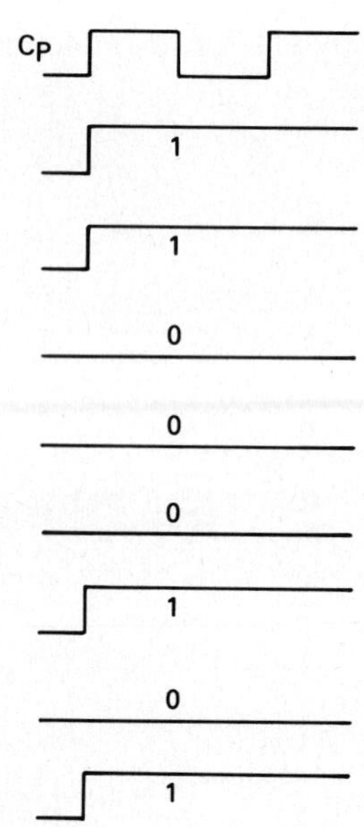

8. only one pair of wires required

9. it is much faster

10.

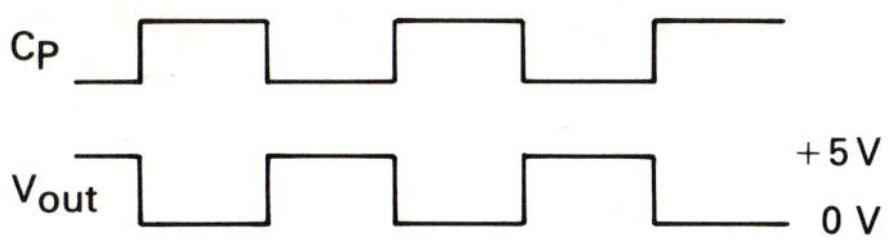

11.

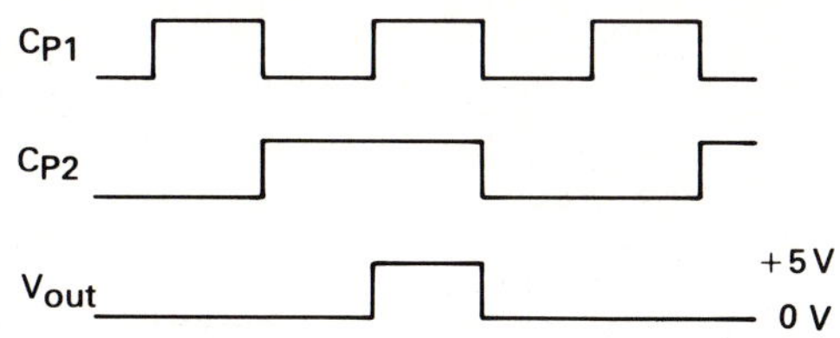

12. D1 forward and D2 reverse
13. +4.3 V
14. 0 V
15. +10 V
16. OFF
17. $V_c = 0$ V
18. faster, quieter, smaller and consumes less power
19. 0 V
20. +12 V
21.

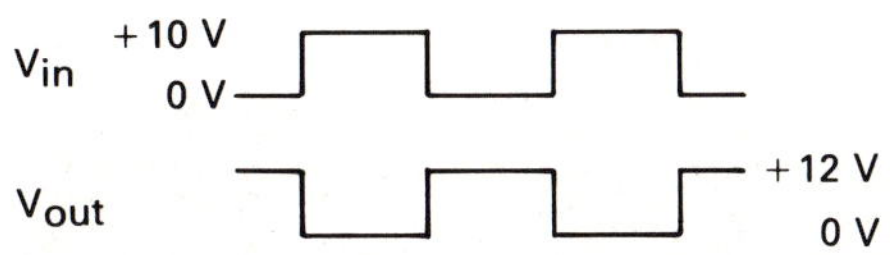

22. The output is the opposite (inverse) of the input.
23. It provides active pull-up and active pull-down.
24. less power
25. switches faster

ANSWERS TO TEST 3

1.

A	B	C	X
0	0	0	0
0	0	1	0
0	1	0	0
0	1	1	0
1	0	0	0
1	0	1	0
1	1	0	0
1	1	1	1

2.

A	B	C	D	Y
0	0	0	0	0
0	0	0	1	1
0	0	1	0	1
0	0	1	1	1
0	1	0	0	1
0	1	0	1	1
0	1	1	0	1
0	1	1	1	1
1	0	0	0	1
1	0	0	1	1
1	0	1	0	1
1	0	1	1	1
1	1	0	0	1
1	1	0	1	1
1	1	1	0	1
1	1	1	1	1

3. $X = A \bullet B \bullet C$
4. $Y = A + B + C + D$
5. 64
6.

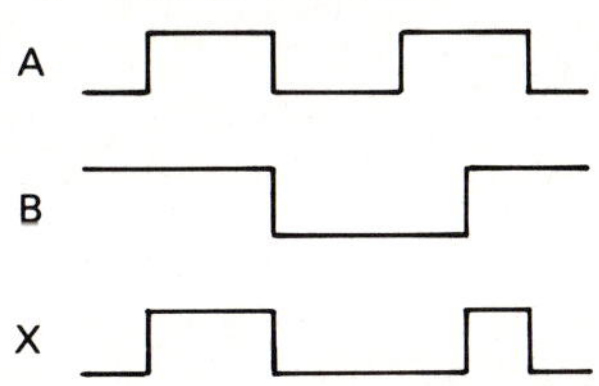

7.

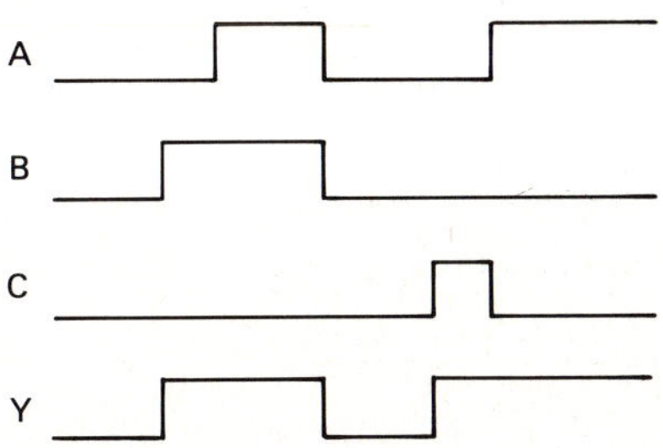

8.

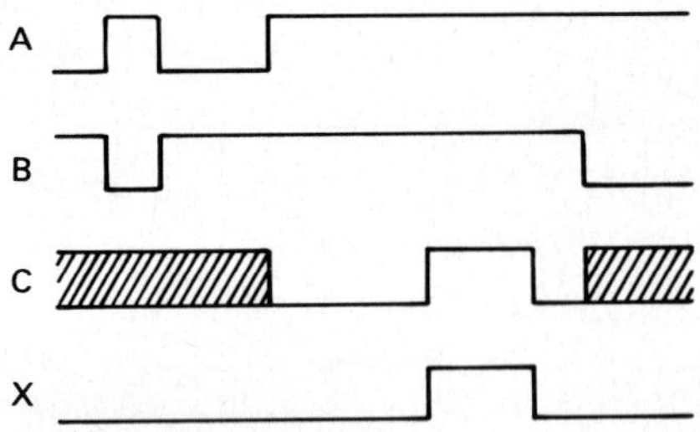

17. The logic probe will indicate, with a light, the high or low status of a pin of a digital IC.
19. With pins 4 and 5 ON and 3 pulsing, output 6 should be pulsing.

9. all four
10. at least one must be low
11. enable must be high 6 usec
12. OR gate
13. active high enable
14. 3
15. 4
16. 3, 6, 8 and 11
18. The logic pulser will insert a string of pulses (1s and 0s) on a pin of a digital IC.
20. no problems

ANSWERS TO TEST 4

1.

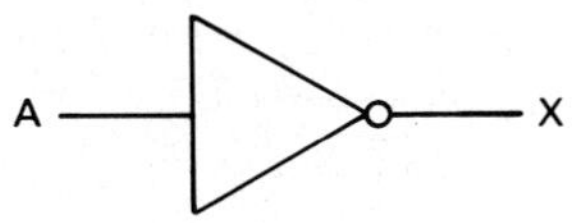

A	X
0	1
1	0

4. $X = \overline{A \bullet B \bullet C}$
5. the output is inverted
7. 16
8.

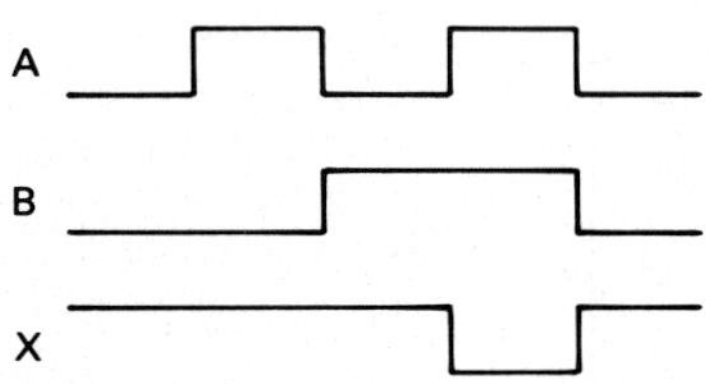

2. Y = C
3.

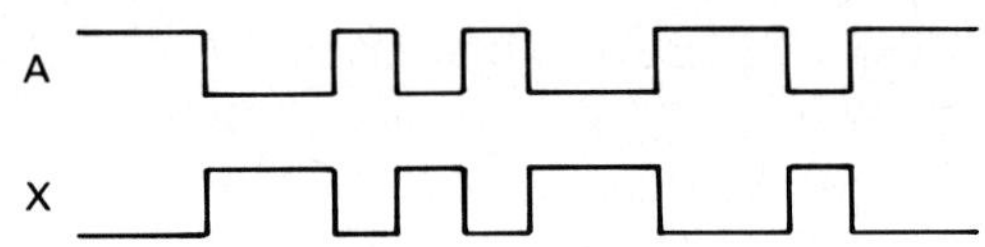

6.

A	B	X
0	0	1
0	1	1
1	0	1
1	1	0

9. OR
11. $Y = \overline{A + B}$
13. A Johnson shift counter
14. $\overline{A} \bullet B$
15.

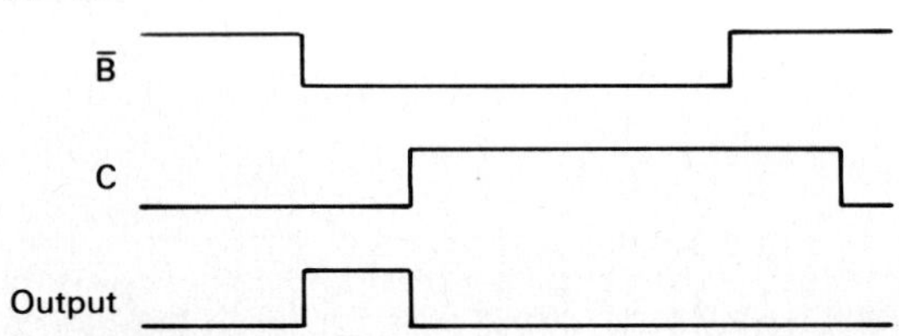

16. inversion
18. 4 circuits
20. 7400 has 4 2-input NAND gates; 7410 has 3 3-input NAND gates

10. A = 0, B = 0, C = 0
12.

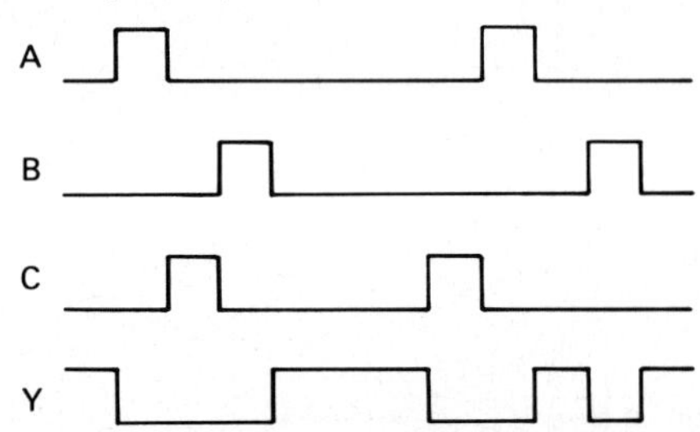

17. 7408
19. 16

ANSWERS TO TEST 5

1. $X = \overline{C} \bullet (A + B) + B \bullet C$
3.

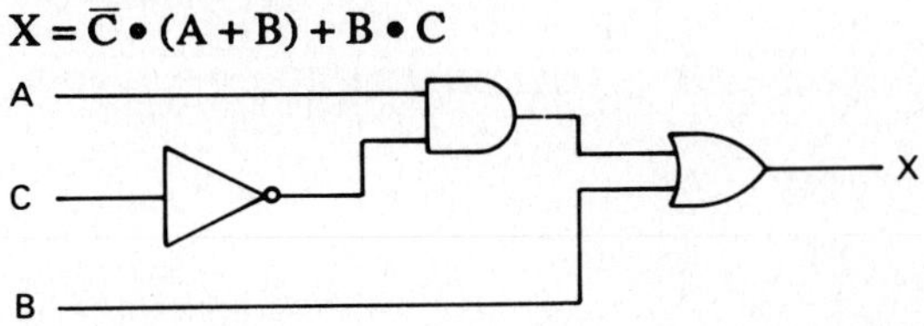

2. $X = A \bullet \overline{C} + B$
4.

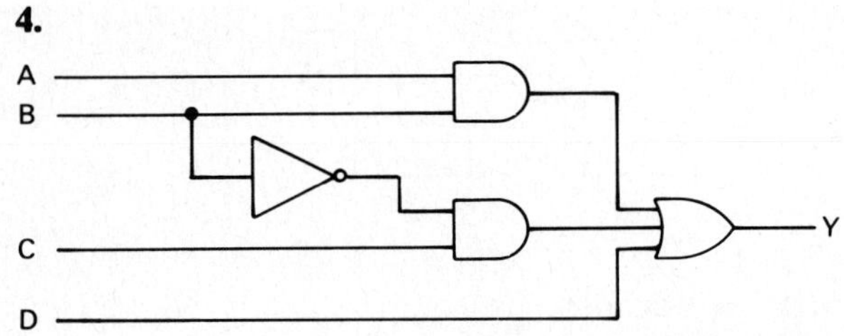

5.

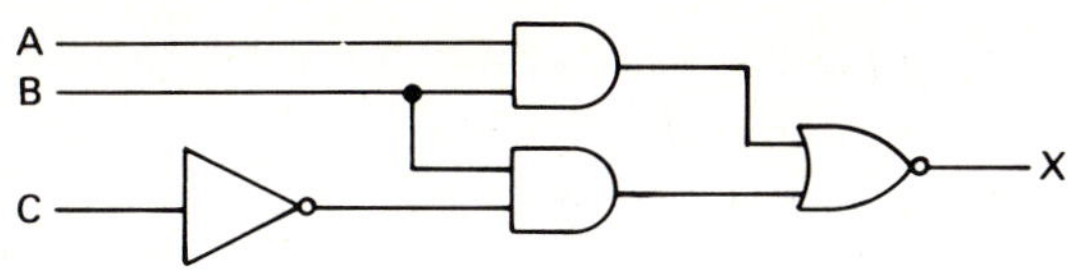

6. $Y = \overline{\overline{A \bullet B} \bullet \overline{B + C}}$

7. $Y = A \bullet B + C$

8.

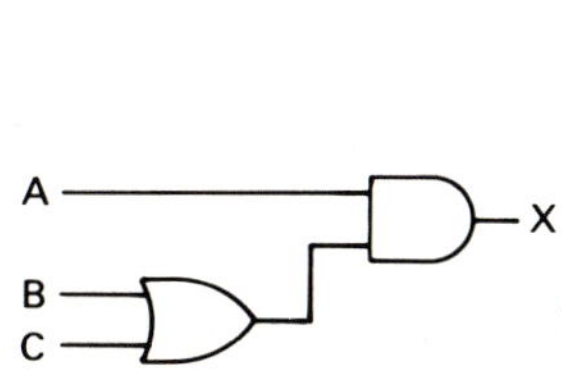

same as

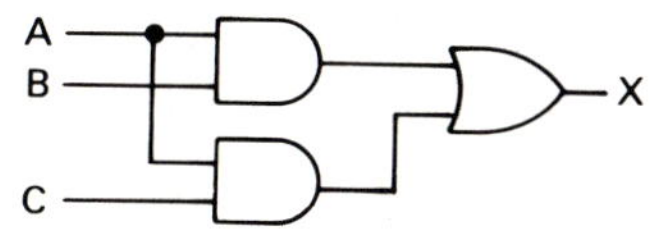

9. $Y = C(\overline{A}B + A\overline{B})$

10. $X = AB\overline{C}$

11.

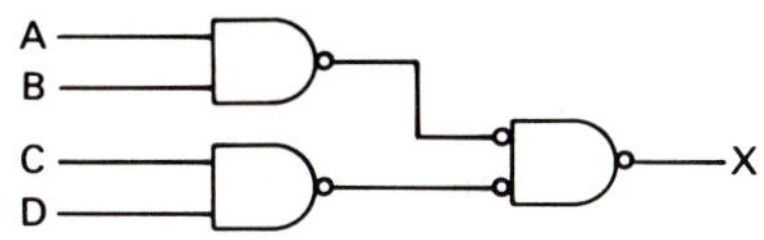

12. $\overline{\overline{\overline{A \bullet B}} \bullet \overline{C} \bullet C} = \overline{A} \bullet \overline{C} + \overline{B} \bullet \overline{C}$

13.

A	B	C	X
0	0	0	0
0	0	1	0
0	1	0	0
0	1	1	0
1	0	0	1
1	0	1	0
1	1	0	0
1	1	1	1

14. $\overline{A} \bullet \overline{B} \bullet \overline{C} + \overline{A} \bullet B \bullet \overline{C} + A \bullet \overline{B}$
$\overline{C} + A \bullet \overline{B} \bullet C + A \bullet B \bullet \overline{C} = Y$

15. $A \bullet \overline{B} + \overline{C} = Y$

16.

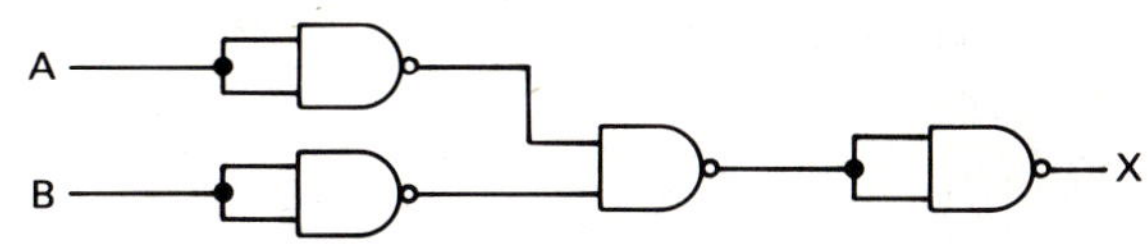

17.

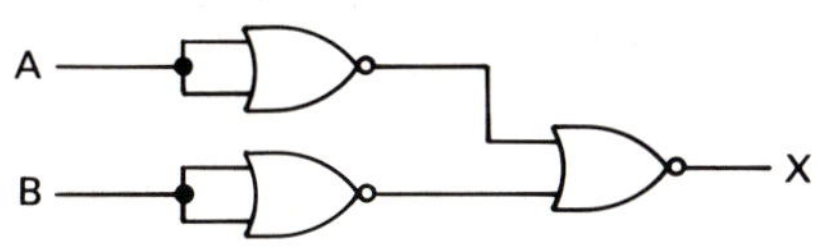

18. $Y = \overline{A}\overline{B}\overline{C} + BCD + \overline{A}BC$

19.

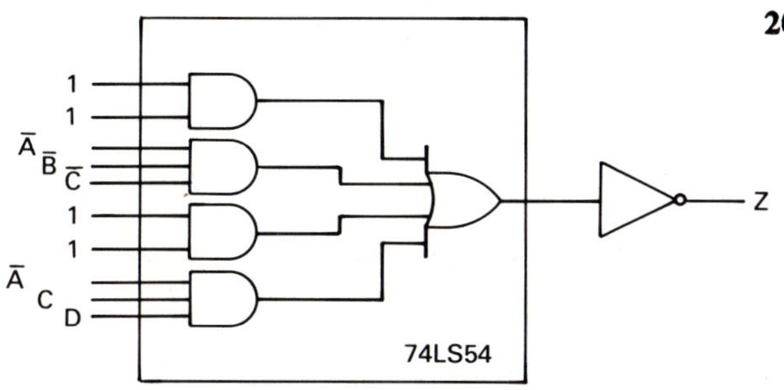

20. $W = AB + BC + A\overline{C} + \overline{B}\overline{C}$

ANSWERS TO TEST 6

1.

A	B	X
0	0	0
0	1	1
1	0	1
1	1	0

2.

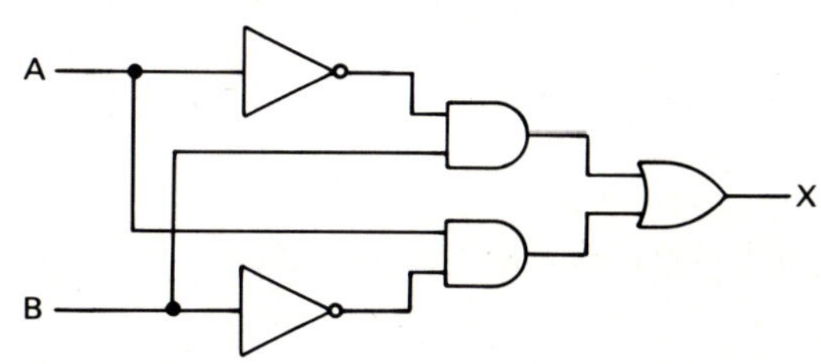

3.

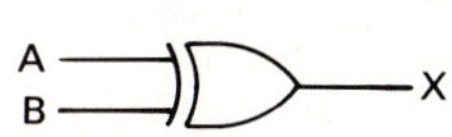

4. 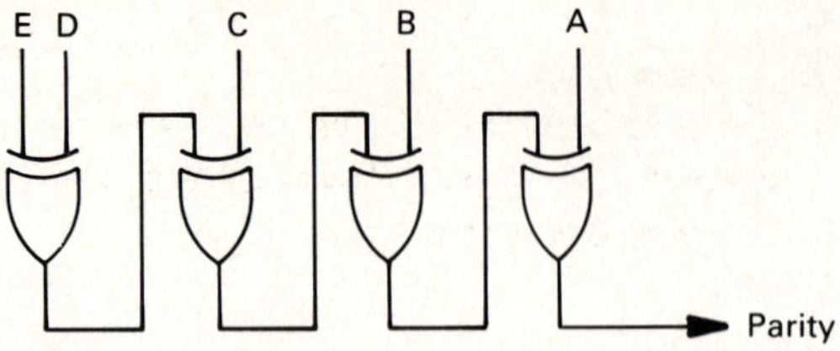

5. Using A as the control, when A = 0, X is the same as B. When A = 1, X is the inverse of B.

6. P = 0, P = 0, P = 1

7. 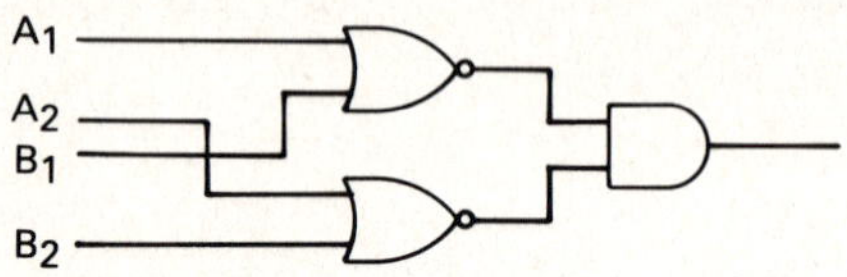

8. The last output is inverted.

9. A checker uses a generator to compare the incoming parity.

10. The output is true if the inputs are equal.

ANSWERS TO TEST 7

1. 0111 1011   0100 0001   0101 1011

2. 0011 0101   0110 1011   1000 0111

3. 0001 1110   0100 1101   0101 1010

4. 0000 0010   0000 0010   0000 0100

5. 11010   11101   00011

6. 0000 0111   1111 1101   1111 0100

7. +5   -4   -8

8.

0000 1101	1111 0110
1111 1001	0000 1111
0000 0110	0000 0101

9.

01001	01100
11101	11001
00110	00101

10.

11101101	00011110	11101100
11111100	11101111	00001010
11101001	00001101	11110110

11. 61   3C   117

12. 22   18   CB

13. 0001 0001   0001 0101   0001 0111

14. Exclusive-OR

15. 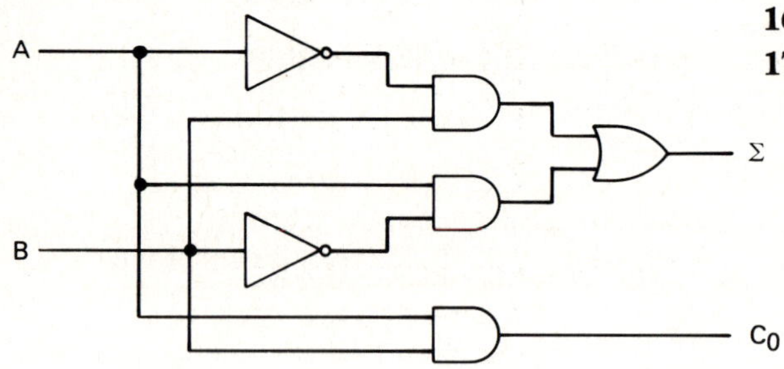

16. YES

17. Full-adder has carry in.

18. 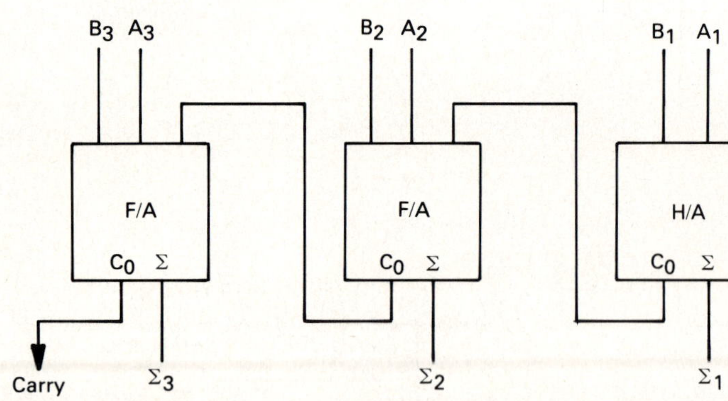

19. to speed-up the circuit

20. Arithmetic and Logic unit - the circuit can not only add and subtract but can do logical functions such as AND, OR and Exclusive-OR.

ANSWERS TO TEST 8

1. 8
2. The magnitude can not only determine if two inputs are equal, but it can also determine if A>B or A<B.
3. A<B means A is less than B. A>B means A is greater than B.
4. A = B = 0, A<B = 0, A>B = 1
5. 4
6. 10
7. a bubble indicates active LOW
8. to turn the circuit on or off
9. 64
10. the output of an encoder is a binary code for a 1-of-N input
11. 10
12. the higher value
13. GS = L, A_0 = H, A_1 = H, A_2 = L, EO = H
14. 000011   000001000
001010   000010100
101000   000101000
110101   011001000
             100001100
15. 2
16. only one digit changes between counts
17. 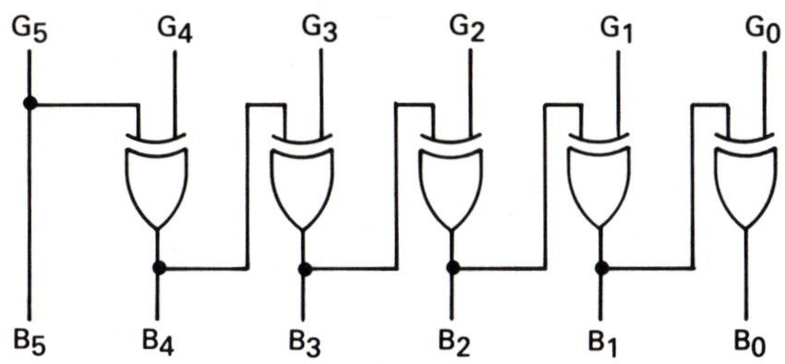

18. 3
19. to activate the entire chip
20. S0 = 1, S1 = 0, S2 = 1
21. 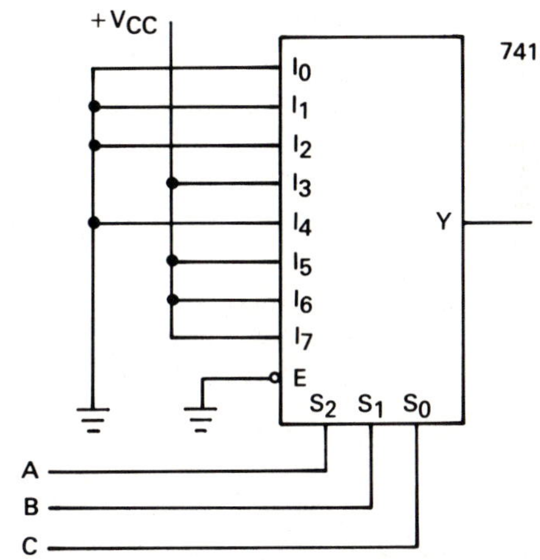

22. the input will be distributed to one of the outputs
23. 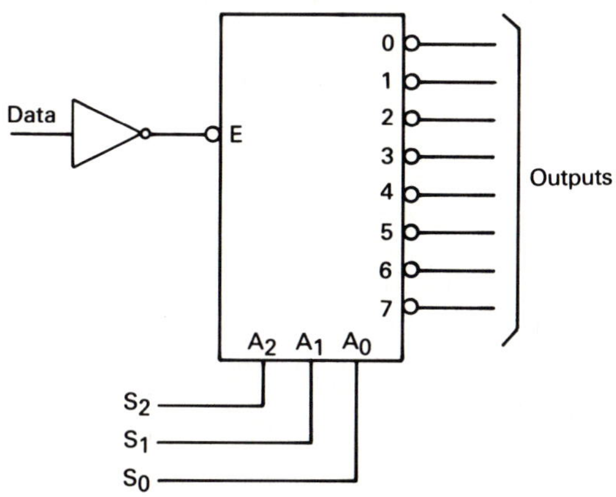

24. The CMOS uses bidirectional switches.

ANSWERS TO TEST 9

1. Q1-OFF, Q2-ON, Q3-OFF, Q4-ON
2. Q1-ON, Q2-OFF, Q3-ON, Q4-OFF
3. 54XX has wider power supply and expanded temperature range
4. high state 8, low state 16
5. 8 mA
6. these are worst case conditions
7. t_{rise} = 4 msec, t_{fall} = 2 msec
8. 10 usec
9. t_{PHL}
10. 82.5mW
11. to provide the high level voltage
12. tie them high
13. to reduce noise, connect between power supply pin and ground
14. less power dissipation
15. TTL
16. can be damaged by static electricity discharge
17. less power dissipation and slower speed
18. The HCT series is input and output voltage compatible with TTL

19. very high speed
21. The propagation delay in nsec is multiplied by the power dissipation in mW.
24. when the supply voltages are different

20. they are simply the inverse of each other
22. to increase the output high level voltage
23. yes
25. 400

ANSWERS TO TEST 10

1. active high
3. no active S or R input
5.

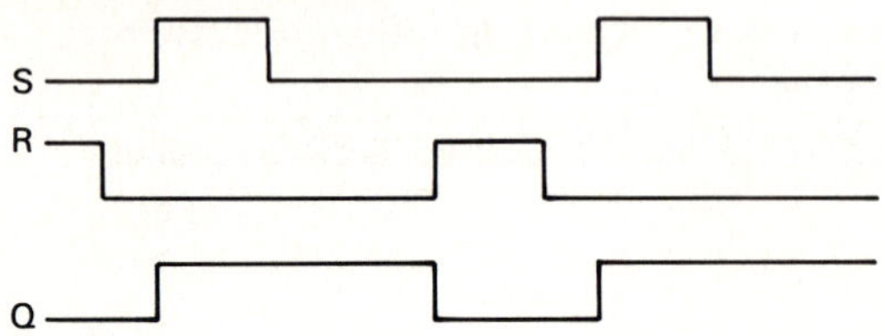

6. they must occur with the gate
7.

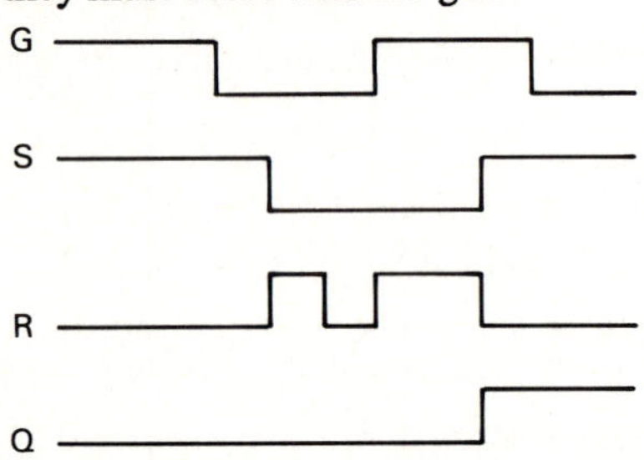

11. 4
12. 7474 is edge-triggered
13.

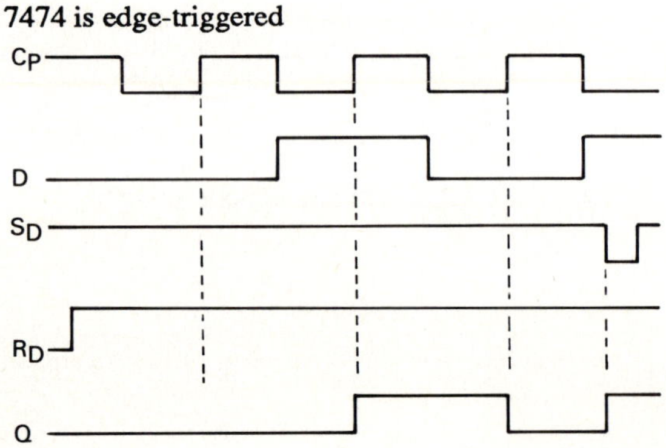

19.

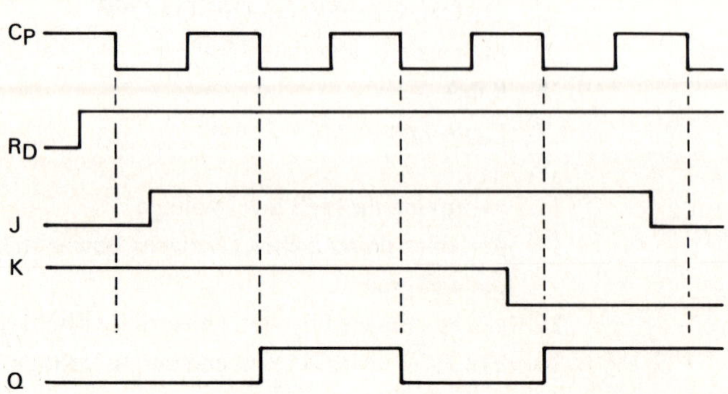

2. invert the S-R inputs
4.

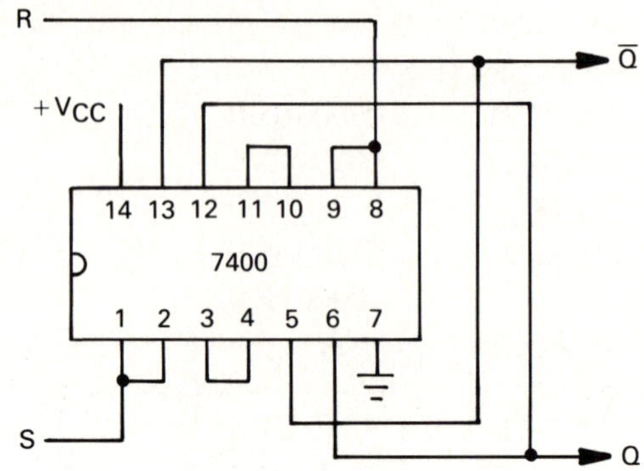

8. when the Q output follows the input
9. The D flip-flop has only a single D input that acts as both a set and reset. There is also no hold condition when the gate is active.
10.

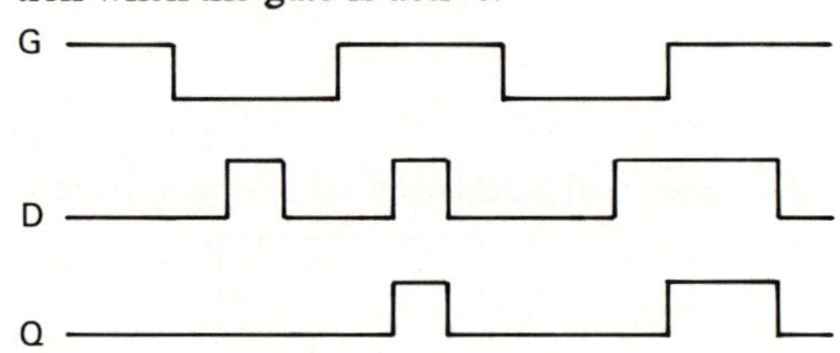

14. J = 0, K = 0
15. J = 1, K = 1
16. when gate is high
17. when gate goes low
18. edge-triggered
20. 7476 is master-slave, 74LS76 is edge triggered

ANSWERS TO TEST 11

1. The inputs to a trigger device are changing at the same time as the active trigger edge.
2. Set up time occurs before the active clock edge, hold time occurs after the active clock edge.
3. NO. $t_{PLH} = 25$ ns, $t_{PHL} = 40$ ns
4. YES
5. worst case condition
6. to provide set up or hold times
7. 5 - 20 ns
8. 20 ns
9. 5 time constants (RC)
10. YES
11. The Schmitt trigger cannot change until the input voltage drops below the lower trip point or goes above the upper trip point.
12. 0.8 volts
13.

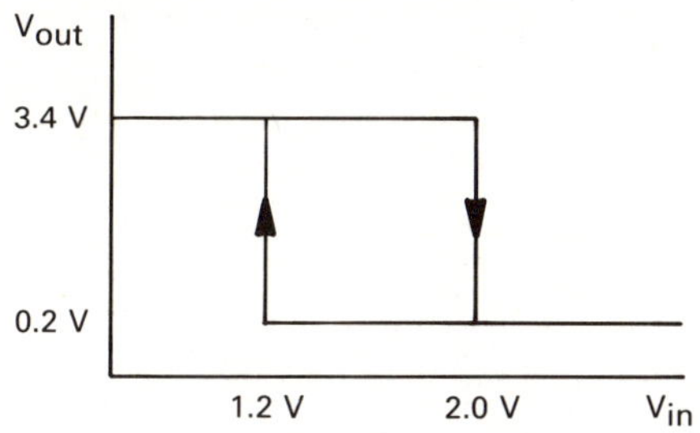

14. switch vibrates ON and OFF
15.

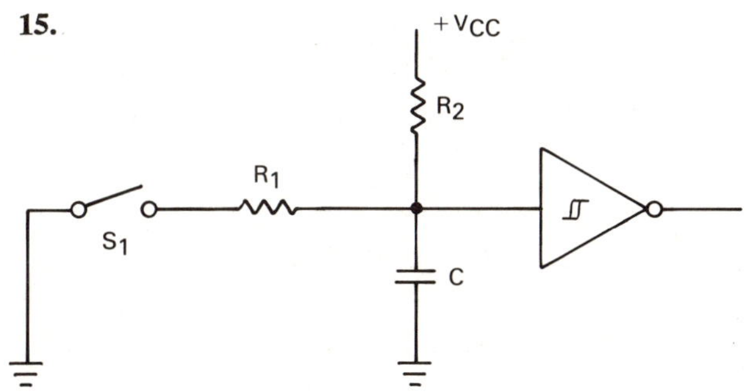

16.

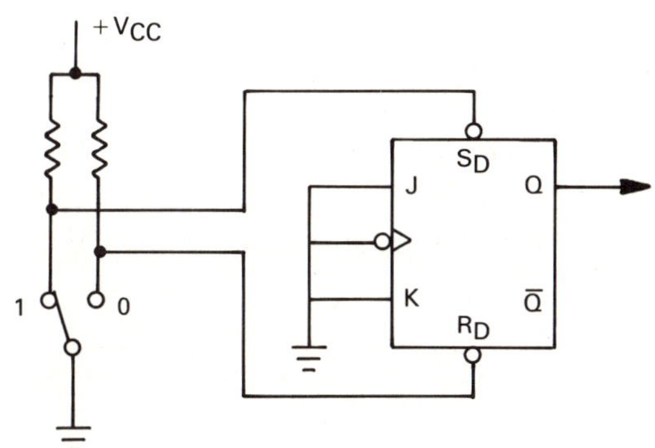

17. +14 VDC
18. to provide a sharp pulse waveform
19. low level current is larger
20. 7.02 mA

ANSWERS TO TEST 12

1. sequential circuits are triggered by timing pulses, combinational circuits are not
2. 16
3. 5
4. 88 ns
5.

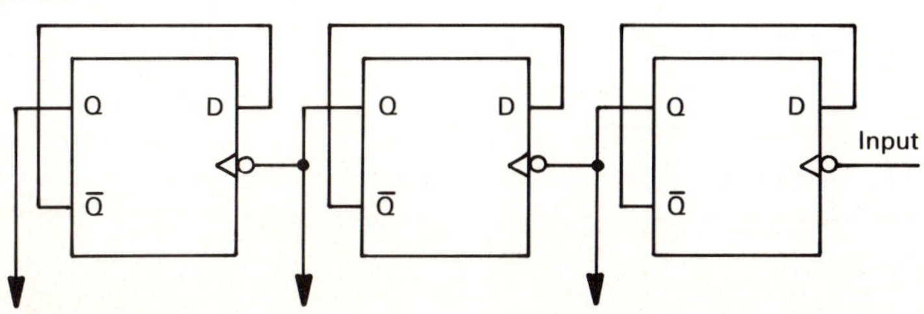

6.

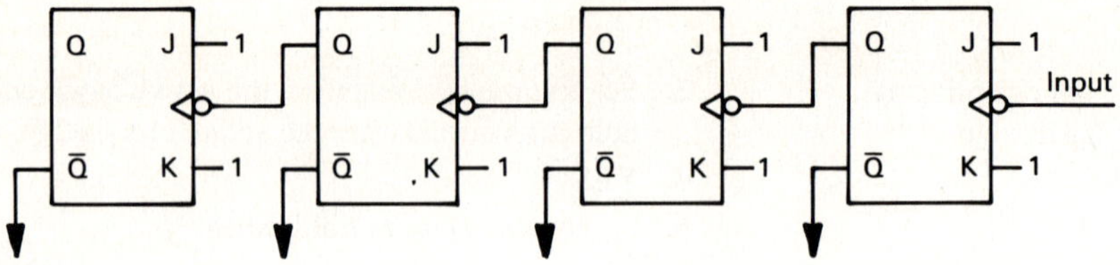

7. 110

8.

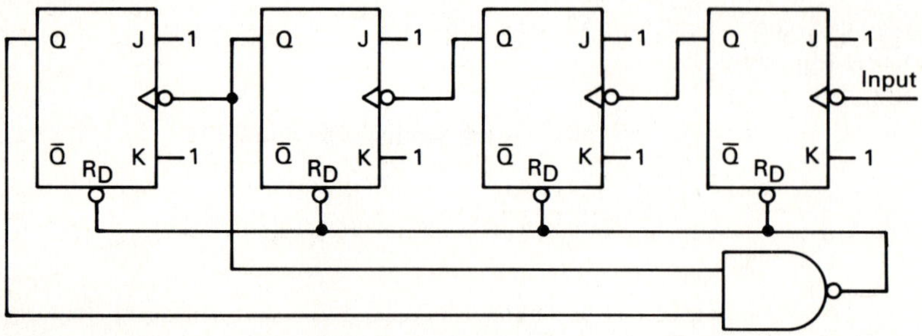

9. Q1 = 11 MHz, Q2 = 5.5 MHz, Q3 = 2.75 MHz, Q4 = 1.375 MHz

10. 12 to 1, 11 to 3, 8 to 2

11. 7490 is MOD-10, 7492 is MOD-12

12. 7490 is MOD-10, 7493 is MOD-16

13. 1 to 12

14. c,d,e,f,g

15. RB0 = 0, a = 0, b = 0, c = 0, d = 1, e = 1, f = 0, g = 0

16. All flip-flops change at the same time.

17.

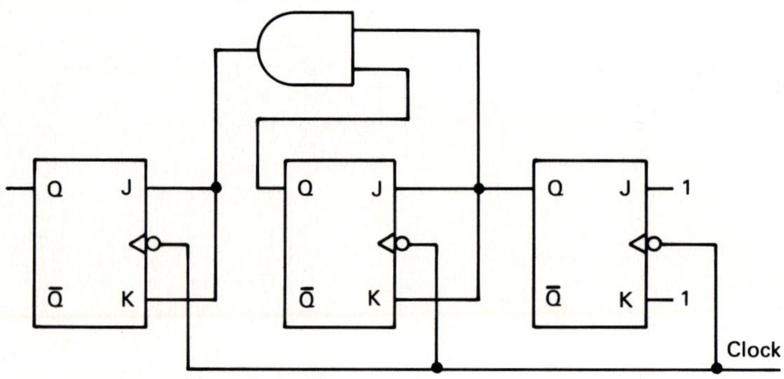

18. ripple counters use less circuits

19.

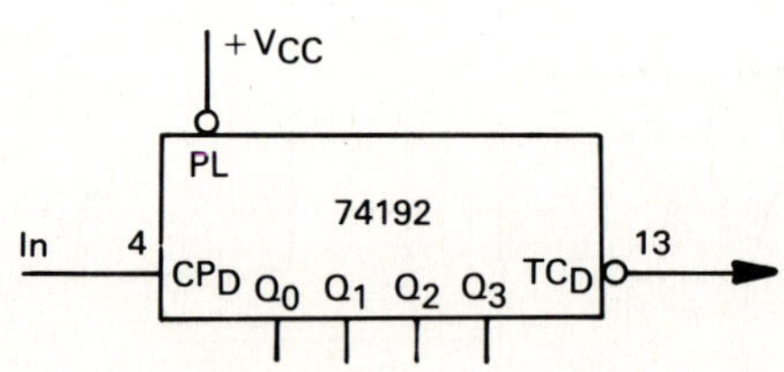

20. all FFs preset with data

ANSWERS TO TEST 13

1. all FFs preset with data

2. 5

3.

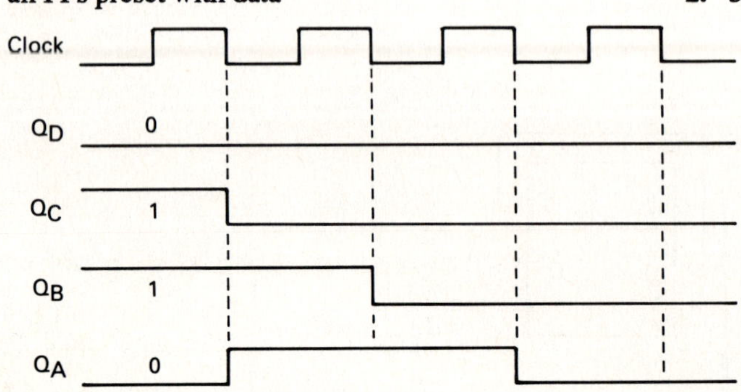

4. the direction of the shift

5. use Q output of each FF

6. serial out connected to serial in

7. control number of clocks

8. a single 1 the rest 0

9.

10. the feedback is reversed

11.

12.

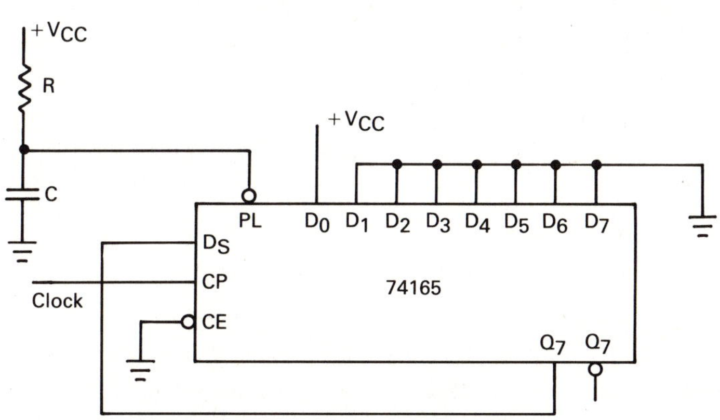

13.

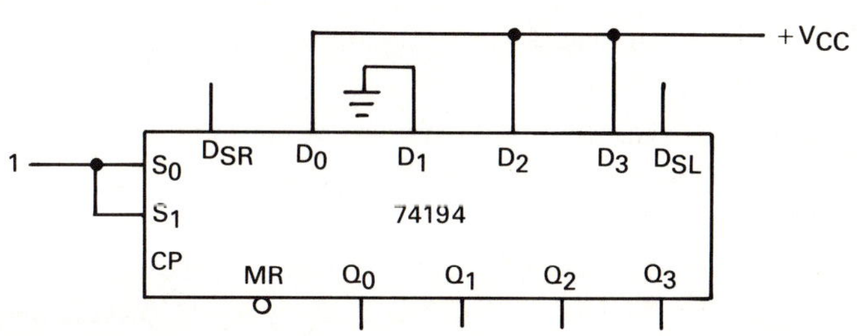

14. HIGH, LOW, float

15. when two or more outputs are connected to the same input

16. activates the three-state buffer

17. to provide the same output that is on the input

18. a set of common connections for transfer of data

19. to catch or remember the output from the microprocessor

20. a buffer that can operate in both directions

ANSWERS TO TEST 14

1. flip-flop
2. astable is free running
3. An RC time constant
4. 0.501 ms
5. 0.309 volts
6. the period of the waveform
7. 34.7 msec
8. R_{int} = 2k C_{ext} = .012 uF
9. the output pulse can be stretched with a retriggerable
10. NO
11. NO
12. 29 usec
13. to compare the input voltages to an internal voltage divider
14. discharge the external capacitor to start the timing over again
15. V_{OH} = 8.5 v, V_{OL} = 0.1 V
16. t_{HI} = 7.95 us, t_{LO} = 6.24 us
17. 56%
18. 21.6%
19. 1.82 ms
20. monostable

ANSWERS TO TEST 15

1. 64
2. 0.234 volts/step
3. I1 = .625 mA, I2 = 1.25 mA
4. 1.875 mA
5. 30 volts
6. too many different values of resistors
7. R-2R ladder
8. 1.5625 volts
9. 4.6875 volts
10. error in the slope of an output staircase waveform
11. offset is when more or less input voltage is required for the first step than what is specified
12. very fast conversion
13. 255
14. 1
15. very slow
16. tests each bit of the digital word starting with the MSB and decides when to keep the bit or reset it to zero
17. 12.8 msec
18. 0.4 msec
19. to be interfaced with a microprocessor
20. to interface a microprocessor to an analog system with a number of different variables

ANSWERS TO TEST 16

1. 11
2. First the address is set, then the data is set on the data bus, and finally, the write pulse stores the data.
3. the time the data must be set before the end of the write pulse
4. dynamic RAM must be refreshed, static RAM does not
5. any address can be accessed in any order
6. 2
7. the refresh cycle keeps the charge on the capacitor cells
8. 2 msec
9. it provides the multiplexing of the address bus and a minimum
10. to keep the number of pins on the chip to a minimum
11. A series of inverter gates with RC circuits between each one
12. mask ROM
13. EPROM
14. EEPROM
15. they do not lose memory when power is removed
16. to activate the three-state buffers on the output
17. 8
18. YES
19. the PAL can program the AND gate inputs, the PLA can program both the AND gate inputs and the OR gate inputs
20. a HAL is programmed at the factory, a PAL is programmed by the user

ANSWERS TO TEST 17

1. 3
2. 8
3. 16
4. DATA BUS
5. c
6. b
7. a
8. a
9. software
10. I/O
11. ALU
12. accumulator
13. IN, OUT
14. Tri-State Buffer
15. Latch
16. machine codes are in binary, mnemonic codes are in shorthand English.
17. assembler
18. English language statements, easier to write, are more inefficient.
19. 1974
20. Intel, Motorola